HANDBOOK OF CHEMICAL REACTIVITY

Kruger Brentt
Publishers

HANDBOOK OF CHEMICAL REACTIVITY

Kahirou Diakite
Josy Furtini

Kruger Brentt
Publishers

2025

Kruger Brentt Publishers UK. LTD.
Company Number 9728962

Regd. Office: 68 St Margarets Road, Edgware, Middlesex HA8 9UU

ISBN: 978-1-78715-294-6

Disclaimer:

Every possible effort has been made to ensure that the information contained in this book is accurate at the time of going to press, and the publisher and author cannot accept responsibility for any errors or omissions, however caused. No responsibility for loss or damage occasioned to any person acting, or refraining from action, as a result of the material in this publication can be accepted by the editor, the publisher or the author. The Publisher is not associated with any product or vendor mentioned in the book. The contents of this work are intended to further general scientific research, understanding and discussion only. Readers should consult with a specialist where appropriate.

Every effort has been made to trace the owners of copyright material used in this book, if any. The author and the publisher will be grateful for any omission brought to their notice for acknowledgement in the future editions of the book.

For information on all our publications visit our website at http://krugerbrentt.com/

PREFACE

"Handbook of Chemical Reactivity" delves into the dynamic world of chemical transformations, offering readers a comprehensive exploration of the factors that govern the reactivity of substances and the pathways by which chemical reactions occur. This book serves as an invaluable resource for students, educators, and researchers in chemistry and related fields, providing a deep understanding of the fundamental principles underlying chemical reactivity.

Chemical reactions are the driving force behind countless natural and synthetic processes, from the metabolism of living organisms to the synthesis of new materials and pharmaceuticals. Understanding the mechanisms and kinetics of chemical reactions is essential for predicting and controlling the behavior of substances in diverse environments and applications.

As editors of this volume, we have endeavored to compile a diverse array of perspectives, methodologies, and case studies from leading experts and practitioners in the field of chemical reactivity. Through a blend of theoretical concepts, experimental techniques, and practical applications, this book covers a wide range of topics, including reaction mechanisms, kinetics, thermodynamics, catalysis, and reaction engineering.

Our overarching goal in presenting this book is to provide readers with a comprehensive and accessible resource that fosters a deeper understanding of chemical reactivity and its implications for research, industry, and everyday life. By demystifying complex concepts, providing real-world examples, and offering practical insights, we aim to empower students, educators, and researchers to explore the fascinating world of chemical reactions with curiosity and confidence.

We extend our sincere gratitude to all the contributors who have generously shared their expertise, experiences, and insights in the field of chemical reactivity. It is our hope that this book will serve as a valuable reference and guide for learners and practitioners alike, inspiring curiosity, critical thinking, and innovation in the study of chemical reactivity and its applications across diverse disciplines.

Kahirou Diakite | Josy Furtini

CONTENTS

1

INTRODUCTION TO CHEMICAL REACTIVITY

DEFINITION AND SIGNIFICANCE OF CHEMICAL REACTIVITY

Chemical reactions are a significant part of technology, pharmaceuticals and life itself. Burning fuels, smelting iron, producing glass and pottery, brewing beer, and making wine and cheese are just a few of the numerous activities that involve chemical reactions that have been recognized and deployed for thousands of years.

A chemical reaction is a process that converts one or more compounds, known as reactants, to one or more distinct substances, known as products. Chemical elements and compounds are both types of substances. A chemical reaction reconfigures the basic atoms of the reactants, resulting in the formation of various substances as products.

Chemical Reactivity

- The reactivity of a material in chemistry is a measure of how easily it conducts a chemical reaction. The material might be involved in the reaction alone or in combination with other atoms or compounds, and it is usually followed by a release of energy.
- The most reactive elements and compounds are capable of spontaneous or explosive ignition. They typically burn in both water and oxygen in the air.
- The temperature influences reactivity. Increasing the temperature makes more energy available for a chemical reaction, which typically increases its likelihood.
- The physical qualities of a sample, its chemical purity, and the presence of additional chemicals all have an impact on its reactivity.
- Simply put, reactivity is determined by the context in which a chemical is observed. Baking soda and water, for example, are not especially reactive,

but baking soda and vinegar easily react to produce carbon dioxide gas and sodium acetate.

- The periodic table's structure enables reactivity predictions. Both strongly electropositive and very electronegative materials have a great proclivity for reaction.
- These elements may be found in the periodic table's top right and lower left corners.
- Now that we can recognize these actors (compounds), we turn to the roles they are inclined to play in the scientific drama staged by the multitude of chemical reactions that define organic chemistry.

We begin by defining some basic terms that will be used frequently as this subject is elaborated.

- **Chemical Reaction:** A transformation resulting in a change of composition, constitution and/or configuration of a compound (referred to as the reactant or substrate).
- **Reactant or Substrate:** The organic compound undergoing change in a chemical reaction. Other compounds may also be involved, and common reactive partners (reagents) may be identified. The reactant is often (but not always) the larger and more complex molecule in the reacting system. Most (or all) of the reactant molecule is normally incorporated as part of the product molecule.
- **Reagent:** A common partner of the reactant in many chemical reactions. It may be organic or inorganic; small or large; gas, liquid or solid. The portion of a reagent that ends up being incorporated in the product may range from all to very little or none.
- **Product(s)** The final form taken by the major reactant(s) of a reaction.
- **Reaction Conditions** The environmental conditions, such as temperature, pressure, catalysts & solvent, under which a reaction progresses optimally. Catalysts are substances that accelerate the rate (velocity) of a chemical reaction without themselves being consumed or appearing as part of the reaction product. Catalysts do not change equilibria positions.

Chemical reactions are commonly written as equations:

$$\textbf{Reactant(s)} \xrightarrow[\left[\text{Reaction Conditions}\right]]{\textbf{Reagent(s)}} \textbf{Product(s)}$$

Significance of Chemical Reactivity

- Chemical reactions are how new forms of matter are made. While nuclear reactions also may produce new matter, nearly all the substances you encounter in daily life are the result of chemical changes.
- Chemical reactions help us understand the properties of matter. By studying the way a sample interacts with other matter, we can learn its chemical properties. These properties can be used to identify an unknown specimen or to predict how different types of matter might react with each other.
- By observing chemical reactions, we are able to understand and explain how the natural world works. Chemical reactions turn food into fuel for your body, make fireworks explode, cause food to change when it is cooked, make soap remove grime, and much more.
- Exciting and entertaining chemical reactions pique interest in science. These types of demonstrations encourage people to ask questions and may lead them into science-related careers.
- Chemical reactions help us to solve crimes and explain mysteries. By analyzing blood and tissue samples, for example, police are able to identify the perpetrators of crimes. Chemical reactions are also the tools we use to date fossils, analyze ancient materials, and better understand how our ancestors lived.
- By observing chemical reactions in outer space, we are able to identify and describe other planets. We can even determine which planets and moons are likely to be able to sustain life.
- Fire, the most important discovery made by human beings, is simply a chemical reaction.
- Without chemical reactions, nothing would ever change. Atoms would stay atoms. New molecules wouldn't form. No organisms could live. If matter didn't react with other matter, the universe would be extremely boring.

HISTORICAL OVERVIEW OF KEY DISCOVERIES IN CHEMICAL REACTIVITY

The B.C. Era

The early years of history didn't have many significant scientific developments, but there was one surprisingly important development in the fifth century B.C.

Democritus (465 B.C.)

First to propose that matter exists in the form of particles. Coined the term 'atoms.' "by convention bitter, by convention sweet, but in reality atoms and void"

1000 to 1600s. From alchemists who started practicing their trade about the year 1000 to the introduction of the first vacuum pump in the mid-1600s, this long period produced a number of scientific developments.

Alchemists (~1000–1650)

Among other things, the alchemists sought a universal solvent, attempted to change lead and other metals into gold, and tried to discover an elixir which would prolong life. The alchemists learned how to use metallic compounds and plant-derived materials to treat diseases.

4 **Chemical Reactivity**

1100s

Oldest written description of lodestone used as a compass.

Sir Robert Boyle (1637–1691)

Formulated the fundamental gas laws. First to propose the combination of small particles to form molecules. Differentiated between compounds and mixtures.

Evangelista Torricelli (1643)

Invented the mercury barometer.

Otto von Guericke (1645)

Constructed the first vacuum pump.

1700s

Scientific discovery stepped up quite a bit in this century, starting with the discovery of oxygen and other gasses to the invention of the electric battery, Benjamin Franklin's experiments with lightning (and his theory about electricity) to theories about the nature of heat.

James Bradley (1728)

Uses aberation of starlight to determine the speed of light to within 5% accuracy. Joseph Priestley (1733–1804)

Discovered oxygen, carbon monoxide, and nitrous oxide. Proposed electrical inverse square law (1767).

C.W. Scheele(1742–1786)

Discovered chlorine, tartaric acid, metal oxidation, and sensitivity of silver compounds to light (photochemistry).

Nicholas Le Blanc (1742–1806)

Invented process for making soda ash from sodium sulfate, limestone, and coal. A.L. Lavoisier (1743–1794)

Discovered nitrogen. Described the composition of many organic compounds. Sometimes regarded as the Father of Chemistry.

A. Volta (1745–1827)

Invented the electric battery.

C.L. Berthollet (1748–1822)

Corrected Lavoiser's theory of acids. Discovered bleaching ability of chlorine. Analyzed combining weights of atoms (stoichiometry).

Edward Jenner (1749-1823)

Development of smallpox vaccine (1776).

Benjamin Franklin (1752)

Demonstrated that lightning is electricity.

John Dalton (1766–1844)

Proposed atomic theory based on measurable masses (1807). Stated law of partial pressure of gasses.

Amedeo Avogadro (1776–1856)

Proposed principle that equal volumes of gasses contain the same number of molecules. Sir Humphry Davy (1778–1829)

Laid foundation of electrochemistry. Studied electrolysis of salts in water. Isolated sodium and potassium.

J.L. Gay-Lussac (1778–1850)

Discovered boron and iodine. Discovered acid-base indicators (litmus). Improved method for making sulfuric acid. Researched behavior of gasses.

J.J. Berzelius (1779–1850)

Classified minerals according to their chemical composition. Discovered and isolated many elements (Se, Th, Si, Ti, Zr). Coined the terms 'isomer' and 'catalyst'.

Charles Coulomb (1795)

Introduced the inverse-square law of electrostatics.

Michael Faraday (1791–1867)

Coined term 'electrolysis'. Developed theories of electrical and mechanical energy, corrosion, batteries, and electrometallurgy. Faraday was not a proponent of atomism.

Count Rumford (1798)

Thought that heat was a form of energy.

Early- to Mid-1800s

The 1800s saw the synthesis of the first organic compound, vulcanization of rubber, the invention of dynamite, the creation of the Periodic Table, the pasteurization of milk and wine, and even the invention of a new way of manufacturing aluminum, among other developments.

F. Wohler (1800–1882)

First synthesis of an organic compound (urea, 1828).

Charles Goodyear (1800–1860)

Discovered vulcanization of rubber (1844). Hancock in England made a parallel discovery.

Thomas Young (1801)

Demonstrated the wave nature of light and the principle of interference. J. von Liebig (1803–1873)

Investigated photosynthesis reaction and soil chemistry. First proposed the use of fertilizers. Discovered chloroform and cyanogen compounds.

Hans Oersted (1820)

Observed that a current in a wire can deflect a compass needle - provided first concrete evidence of the connection between electricity and magnetism.

Thomas Graham (1822–1869)

Studied diffusion of solutions through membranes. Established foundations of colloid chemistry.

Louis Pasteur (1822–1895)

First recognition of bacteria as disease-causing agents. Developed field of immunochemistry. Introduced heat-sterilization of wine and milk (pasteurization). Saw optical isomers (enantiomers) in tartaric acid.

William Sturgeon (1823)

Invented the electromagnet.

Sadi Carnot (1824)

Analyzed heat engines.

Simon Ohm (1826)

Stated law of electrical resistance.

Robert Brown (1827)

Discovered Brownian motion.

Joseph Lister (1827–1912)

Initiated use of antiseptics in surgery, e.g., phenols, carbolic acid, cresols.

A. Kekulé (1829–1896)

Father of aromatic chemistry. Realized four-valent carbon and structure of benzene ring. Predicted isomeric substitutions (ortho-, meta-, para-).

Alfred Nobel (1833–1896)

Invented dynamite, smokeless powder, and blasting gelatin. Established international awards for achievements in chemistry, physics, and medicine (Nobel Prize).

Dmitri Mendeléev (1834–1907)

Discovered periodicity of the elements. Compiled the first Periodic Table with elements arranged into 7 groups (1869).

J.W. Hyatt (1837–1920)

Invented the plastic Celluloid (nitrocellulose modified using camphor)(1869). Sir W.H. Perkin (1838–1907)

Synthesized first organic dye (mauveine, 1856) and first synthetic perfume (coumarin). F.K. Beilstein (1838–1906)

Compiled Handbuchder organischen Chemie, a compendium of the properties and reactions of organics.

Josiah W. Gibbs (1839–1903)

Stated three principal laws of thermodynamics. Described the nature of entropy and established a relation between chemical, electric, and thermal energy.

H. Chardonnet (1839–1924)

Produced a synthetic fiber (nitrocellulose).

James Joule (1843)

Experimentally demonstrated that heat is a form of energy.

L. Boltzmann (1844–1906)

Developed kinetic theory of gasses. Viscosity and diffusion properties are summarized in Boltzmann's Law.

W.K. Roentgen (1845–1923)

Discovered x-radiation (1895). Nobel Prize in 1901.

Lord Kelvin (1838)

Described the absolute zero point of temperature.

James Joule (1849)

Published results from experiments showing that heat is a form of energy. H.L. **Le Chatelier (1850–1936)**

Fundamental research on equilibrium reactions (Le Chatelier's Law), combustion of gasses, and iron and steel metallurgy.

H. Becquerel (1851–1908)

Discovered radioactivity of uranium (1896) and deflection of electrons by magnetic fields and gamma rays. Nobel Prize in 1903 (with the Curies).

H. Moisson (1852–1907)

Developed electric furnace for making carbides and purifying metals. Isolated fluorine (1886). Nobel Prize in 1906.

Emil Fischer (1852–1919)

Studied sugars, purines, ammonia, uric acid, enzymes, nitric acid. Pioneer research in sterochemistry. Nobel Prize in 1902.

Sir J.J. Thomson (1856–1940)

Research on cathode rays proved existence of electrons (1896). Nobel Prize in 1906. J. Plucker (1859)

Built one of the first gas discharge tubes (cathode ray tubes).

James Clerk Maxwell (1859)

Described the mathematical distribution of the velocities of molecules of a gas. Svante Arrhenius (1859–1927)

Researched rates of reaction versus temperature (Arrhenius equation) and electrolytic dissociation. Nobel Prize in 1903.

Hall, Charles Martin (1863–1914)

Invented method of manufacturing aluminum by the electrochemical reduction of alumina. Parallel discovery by Heroult in France.

Late 1800s-1900s

From the development of the first synthetic resin to discoveries about the nature of radiation and the development of penicillin, this period produced many scientific milestones.

Leo H. Baekeland (1863–1944)

Invented phenolformaldehyde plastic (1907). Bakelite was the first completely synthetic resin.

Walther Hermann Nernst (1864–1941)

Nobel Prize in 1920 for work in thermochemistry. Performed basic research in electrochemistry and thermodynamics.

A. Werner (1866–1919)

Introduced concept of coordination theory of valence (complex chemistry). Nobel Prize in 1913.

Marie Curie (1867–1934)

With Pierre Curie, discovered and isolated radium and polonium (1898). Studied radioactivity of uranium. Nobel Prize in 1903 (with Becquerel) in physics; in chemistry 1911.

F. Haber (1868–1924)

Synthesized ammonia from nitrogen and hydrogen, the first industrial fixation of atmospheric nitrogen (the process was further developed by Bosch). Nobel Prize 1918.

Lord Kelvin (1874)

Stated the second law of thermodynamics.

Sir Ernest Rutherford (1871–1937)

Discovered that uranium radiation is composed of positively charged 'alpha' particles and negatively charged 'beta' particles (1989/1899). First to prove radioactive decay of heavy elements and to perform a transmutation reaction (1919). Discovered half life of radioactive elements. Established that the nucleus was small, dense, and positively charged. Assumed that electrons were outside the nucleus. Nobel Prize in 1908.

James Clerk Maxwell (1873)

Proposed that electric and magnetic fields filled space.

G.J. Stoney (1874)

Proposed that electricity consisted of discrete negative particles he named 'electrons'.

Gilbert N. Lewis (1875–1946)

Proposed electron-pair theory of acids and bases.

F.W. Aston (1877–1945)

Pioneer research on isotope separation by mass spectrograph. Nobel Prize 1922.

Sir William Crookes (1879)

Discovered that cathode rays travel in straight lines, impart a negative charge, are deflected by electric and magnetic fields (indicating negative charge), cause glass to fluoresce, and cause pinwheels in their path to spin (indicating mass).

Hans Fischer (1881–1945)

Research on porphyrins, chlorophyll, carotene. Synthesized hemin. Nobel Prize in 1930.

Irving Langmuir (1881–1957)

Research in the fields of surface chemistry, monomolecular films, emulsion chemistry, electric discharges in gasses, cloud seeding. Nobel Prize in 1932.

Hermann Staudinger (1881–1965)

Studied high-polymer structure, catalytic synthesis, polymerization mechanisms. Nobel Prize in 1963.

Sir Alexander Flemming (1881–1955)

Discovered the antibiotic penicillin (1928). Nobel Prize in 1945.

E. Goldstein (1886)

Used cathode ray tube to study 'canal rays', which possessed electrical and magnetic properties opposite those an electron.

Heinrich Hertz (1887)

Discovered the photoelectric effect.

Henry G.J. Moseley (1887–1915)

Discovered the relation between the frequency of the x-rays emitted by an element and its atomic number (1914). His work led to the reorganization of the periodic table based on an atomic number rather than atomic mass.

Heinrich Hertz (1888)

Discovered radio waves.

Roger Adams (1889–1971)

Industrial research on catalysis and methods of structural analysis. Thomas Midgley (1889–1944)

Discovered tetraethyl lead and it used as an antiknock treatment for gasoline (1921). Discovered fluorocarbon refrigerants. Performed early research on synthetic rubber.

Vladimir N. Ipatieff (1890?–1952)

Research and development of catalytic alkylation and isomerisation of hydrocarbons (together with Herman Pines).

Sir Frederick Banting (1891–1941)

Isolated the insulin molecule. Nobel Prize in 1923.

Sir James Chadwick (1891–1974)

Discovered the neutron (1932). Nobel Prize in 1935.

Harold C. Urey (1894-1981)

One of the leaders of the Manhattan Project. Discovered deuterium. Nobel Prize 1934.

Wilhelm Roentgen (1895)

Discovered that certain chemicals near a cathode ray tube glowed. Found highly penetrating rays that were not deflected by a magnetic field, which he named 'x rays'.

Henri Becquerel (1896)

While studying effects of x-rays on photographic film, he discovered that some chemicals spontaneously decompose and emit very penetrating rays.

Wallace Carothers (1896–1937)

Synthesized neoprene (polychloroprene) and nylon (polyamide).

Thomson, Joseph J. (1897)

Discovered the electron. Used a cathode ray tube to experimentally determine the charge to mass ratio of an electron. Found that 'canal rays' were associated with the proton H^+.

Plank, Max (1900)

Stated radiation law and Planck's constant.

Soddy (1900)

Observed spontaneous disintegration of radioactive elements into 'isotopes' or new elements, described 'half-life', made calculations of the energy of decay.

George B. Kistiakowsky (1900–1982)

Devised the detonating device used in the first atomic bomb.

Werner K. Heisenberg (1901–1976)

Developed the orbital theory of chemical bonding. Described atoms using a formula related to the frequencies of spectral lines. Stated the Uncertainty Principle (1927). Nobel Prize in 1932.

Enrico Fermi (1901-1954)

First to achieve a controlled nuclear fission reaction (1939/1942). Performed fundamental research on subatomic particles. Nobel Prize in 1938.

Nagaoka (1903)

Postulated a 'Saturnian' atom model with flat rings of electrons revolving about a positively charged particle.

Abegg (1904)

Discovered that inert gasses have a stable electron configuration which results in their chemical inactivity.

Hans Geiger (1906)

Developed an electrical device which made an audible 'click' when hit with alpha particles.

Ernest O. Lawrence (1901–1958)

Invented the cyclotron, which was used to create the first synthetic elements. Nobel Prize in 1939.

Wilard F. Libby (1908–1980)

Developed carbon-14 dating technique. Nobel Prize in 1960.

Ernest Rutherford and Thomas Royds (1909)

Demonstrated that alpha particles are doubly ionized helium atoms.

Niels Bohr (1913)

Devised quantum model of the atom in which atoms had orbital shells of electrons.

Robert Milliken (1913)

Experimentally determined the charge and mass of an electron using an oil drop.

F.H.C Crick (1916–2004) with James D. Watson

Described the structure of the DNA molecule (1953).

Robert W. Woodward (1917-1979)

Synthesized many compounds, including cholesterol, quinine, chlorophyll, and cobalamin. Nobel Prize in 1965.

F.W. Aston (1919)

Use a mass spectrograph to demonstrate the existence of isotopes.

Louis de Broglie (1923)

Described the particle/wave duality of electrons.

Werner Heisenberg (1927)

Stated the quantum uncertainty principle. Described atoms using a formula based on the frequencies of spectral lines.

John Cockcroft, Ernest Walton (1929)

Constructed a linear accelerator and bombarded lithium with protons to produce alpha particles.

Erwin Schodinger (1930)

Described electrons as continuous clouds. Introduced 'wave mechanics' to mathematically describe the atom.

Paul Dirac (1930)

Proposed anti-particles and discovered the anti-electron (positron) in 1932. (Segre/ Chamberlain detected the anti-proton in 1955).

James Chadwick (1932)

Discovered the neutron.

Carl Anderson (1932)

Discovered the positron.

Wolfgang Pauli (1933)

Proposed the existence of neutrinos as a means of accounting for what appeared to be a violation of the law of conservation of energy in some nuclear reactions.

Enrico Fermi (1934)

Formulated his theory of beta decay.

Lise Meitner, Otto Hahn, Fritz Strassmann (1938)

Verified that heavy elements capture neutrons to form fisionable unstable products in a process which ejects more neutrons, thus continuing the chain reaction. that heavy elements capture neutrons to form fisionable unstable products in a process which ejects more neutrons, thus continuing the chain reaction.

Glenn Seaborg (1941–1951)

Synthesized several transuranium elements and suggested a revision to the layout of the periodic table.

ROLE OF CHEMICAL REACTIVITY IN CHEMICAL TRANSFORMATIONS

Reactivity plays a crucial role in chemical transformations, dictating the speed and outcome of various reactions. It is the fundamental property that governs how readily a substance undergoes a chemical change when exposed to different conditions. Understanding reactivity is essential for chemists as it allows them to predict and control the behavior of substances in a wide range of applications, from drug synthesis to environmental remediation. Reactivity can be viewed from different perspectives, including electronic, steric, and thermodynamic considerations, each providing valuable insights into the intricacies of chemical reactions.

- **Electronic Factors:** The electronic structure of molecules heavily influences their reactivity. Substances with high electron density or those containing highly reactive functional groups are more prone to undergo chemical transformations. For example, alkenes possess a double bond that provides a region of high electron density, making them highly reactive towards addition reactions. On the other hand, aromatic compounds exhibit exceptional stability due to delocalized - electrons, resulting in lower reactivity compared to alkenes.
- **Steric Effects:** Steric hindrance refers to the spatial arrangement of atoms within a molecule that can impede or facilitate chemical reactions. Bulky substituents or crowded environments around reactive sites can hinder the approach of other molecules or reagents, reducing reactivity. For instance, tertiary alkyl halides often exhibit slower reaction rates compared to primary alkyl halides due to increased steric hindrance around the carbon atom bearing the leaving group.
- **Thermodynamics:** Thermodynamic factors also influence reactivity by determining whether a reaction is energetically favorable or not. The difference in energy between reactants and products, known as the Gibbs free energy change (G), determines the spontaneity of a reaction. Reactions with negative G values are exergonic and tend to proceed spontaneously, while those with positive G values require an input of energy and are endergonic. For example, the combustion of methane (CH4) with oxygen (O2) has a highly negative G value, indicating its high reactivity and ability to release energy.

A chemical reaction is a process that leads to the chemical transformation of one set of chemical substances to another. When chemical reactions occur, the atoms are rearranged and the reaction is accompanied by an energy change as new products are generated. Classically, chemical reactions encompass changes that only involve the positions of electrons in the forming and breaking of chemical bonds between atoms, with no change to the nuclei (no change to the elements present), and can often be described by a chemical equation. Nuclear chemistry is a sub-discipline of chemistry that involves the chemical reactions of unstable and radioactive elements where both electronic and nuclear changes can occur.

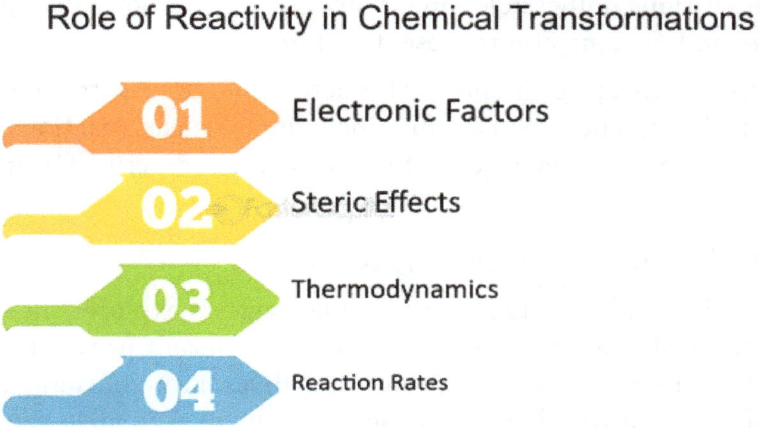

Fig. 1.1 Role of Chemical Reactivity in Chemical Transformations

The substance (or substances) initially involved in a chemical reaction are called reactants or reagents. Chemical reactions are usually characterized by a chemical change, and they yield one or more products, which usually have properties different from the reactants. Reactions often consist of a sequence of individual sub-steps, the so-called elementary reactions, and the information on the precise course of action is part of the reaction mechanism. Chemical reactions are described with chemical equations, which symbolically present the starting materials, end products, and sometimes intermediate products and reaction conditions.

Chemical reactions happen at a characteristic reaction rate at a given temperature and chemical concentration. Some reactions produce heat and are called exothermic reactions, while others may require heat to enable the reaction to occur, which are called endothermic reactions. Typically, reaction rates increase with increasing temperature because there is more thermal energy available to reach the activation energy necessary for breaking bonds between atoms.

A reaction may be classified as redox in which oxidation and reduction occur or non-redox in which there is no oxidation and reduction occurring. Most simple redox reactions may be classified as a combination, decomposition, or single displacement reaction.

Different chemical reactions are used during chemical synthesis in order to obtain the desired product. In biochemistry, a consecutive series of chemical reactions (where the product of one reaction is the reactant of the next reaction) form metabolic pathways. These reactions are often catalyzed by protein enzymes. Enzymes increase the rates of biochemical reactions, so that metabolic syntheses and decompositions impossible under ordinary conditions can occur at the temperature and concentrations present within a cell.

The general concept of a chemical reaction has been extended to reactions between entities smaller than atoms, including nuclear reactions, radioactive decays and reactions between elementary particles, as described by quantum field theory.

TYPES OF CHEMICAL REACTIONS

Types of Chemical Reactions as the name suggests explores the classification of various types of chemical reactions based on different parameters. As we are surrounded by chemical reactions in our daily lives. From burning a match stick to making food all of the things and events in our lives involve various types of chemical reactions. Understanding the different types of chemical reactions is key to understanding how matter changes and interacts with one another. From combustion reactions that power our cars and homes, to acid-base reactions that occur in our bodies and the natural world, there are many different types of chemical reactions with a wide range of applications

Combination, decomposition, single-replacement, double-replacement, redox, combustion, precipitation, neutralization, etc. are some of the types of chemical reactions using various parameters as the basis. In this article, we will learn about it all the types of chemical reactions and the various bases for these classifications of chemical reactions as well.

Based on the different properties, types of change or reactants and products, we can classify chemical reactions. Some of these classifications on various basis are as follows:

Based on Product and Reactant

Based on the formation of different kinds of products or change in the condition of reactants, different types of chemical reactions are:

- Combustion reaction
- Neutralization reaction
- Decomposition reaction
- Redox Reaction

- Double-Displacement Reaction

Combination Reaction

In the Combination Reaction, two or more reactants combine together to form a single product. As in this reaction, new compounds are formed with the help of other compounds, so it is also called a synthesis reaction.

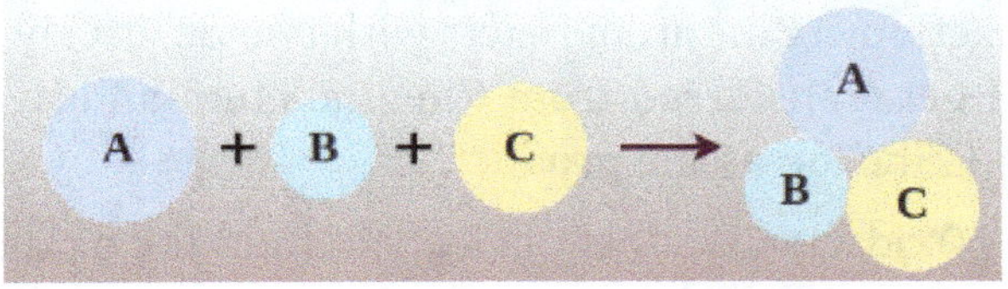

Some Examples of combination reactions are as follows:

- Reaction of quick lime (Calcium Oxide) with water, results in calcium hydroxide, $CaO(s)+H_2O(l)\rightarrow Ca(OH)_2(aq)$
- Reaction of sulfur trioxide gas with water results in sulfuric acid, $SO_3(g)+H_2O(l)\rightarrow H_2SO_4(aq)$

Decomposition Reaction

In the Decomposition reaction, a single component breaks down into two or more products, to break the bonds between different components a lot of heat is required, we can provide it in the form of heat, electricity or sunlight, etc.

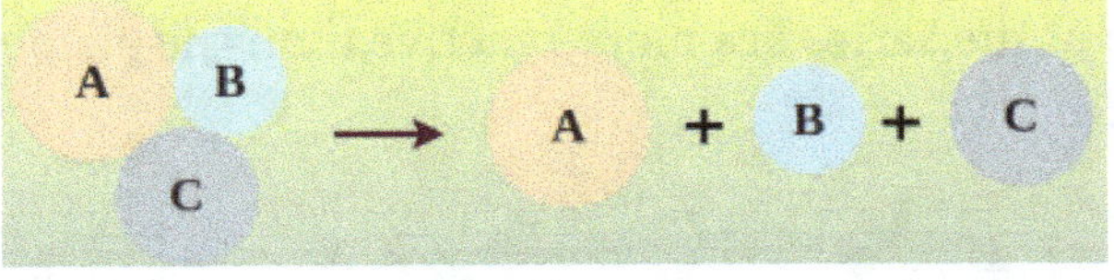

Some examples of decomposition reactions are as follows:

- Decomposition of carbonic acid, $H2CO_3(aq) \rightarrow H_2O(l) + CO_2(g)$
- Decomposition of Nickel (III) Oxide, $2Ni_2O_3(aq) \rightarrow 4Ni(s) + 3O_2(g)$
- The decomposition of calcium carbonate (CaCo3) gives CaO (Quick Lime) which is a major component of cement. $CaCO_3(s)$ heat > $CaO(s) + CO2(g)$
- Calcium hydroxide (Slaked lime) on decomposition gives Quick lime (CaO). $Ca(OH)_2$ heat> $CaO(s) + H_2O\ (l)$

Redox Reaction

Redox Reaction which is also called Reduction-oxidation reaction is the reaction, in which reduction(gain of the electron) and oxidation (loss of electron) takes place simultaneously. So in other words, it is the process of transfer of electrons between different elements and compounds to form new compounds.

Consider an example of an electron-chemical cell-like redox reaction between Zinc and Hydrogen.

$$Zn + 2H+ \rightarrow Zn^{2}+H_2$$

For example, take the reaction of Iron with copper sulfate,

$$CuSO4\ (aq) + Fe(s) \rightarrow FeSO_4\ (aq) + Cu(s)$$

In the above reaction, a Zinc atom reacts with two positively charged hydrogen to which electrons get transferred from the zinc atom and hydrogen becomes a stable molecule and Zinc ion as the product.

- **Oxidation** - Oxidation is a loss of electrons and the oxidation state of given reactant increases. In the above reaction, copper goes through oxidation as it gains electrons and is a reducing agent in this example.
- **Reduction** - Reduction is a gain of electrons and the reactant state of a given reactant decreases. In the above reaction, iron goes through reduction as it losses electrons and is an oxidizing agent in this example.

Single Displacement Reaction

In a Single Displacement Reaction, more reactive metal displaces less reactive metal from its salt or other compounds. In these reactions, through reactivity series, products can be determined. Reactivity series is a series of elements in which elements are arranged in decreasing order of their reactivity. Thus, the elements present at the top of the series are more reactive than the elements present at the bottom of the series.

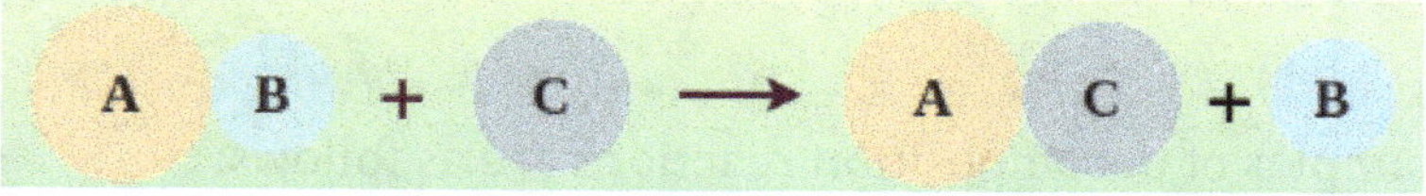

For example, take the reaction between potassium and magnesium chloride (MgCl2). 2K + MgCl2 → 2KCl + Mg

In this reaction, potassium displaces magnesium from its salt because of the more reactivity of potassium than magnesium. Potassium is present at the top of the reactivity series and hence, it is the most reactive element.

Some more examples of Single Displacement Reactions include:

- Zinc reacting with hydrochloric acid to produce zinc chloride and hydrogen gas: $Zn + 2HCl \rightarrow ZnCl_2 + H_2$
- Iron reacting with copper sulfate to produce copper and iron sulfate: $Fe + CuSO_4 \rightarrow Cu + FeSO_4$
- Zinc reacting with silver nitrate to produce zinc nitrate and silver: $Zn + 2AgNO_3 \rightarrow Zn(NO_3)_2 + 2Ag$

Double-Displacement Reaction

It is a kind of displacement reaction in which two chemical species react and consequently, their ions i.e. cations and anions switch places forming two new products. now look at the reaction between silver nitrate and sodium chloride. The products of the reaction will be silver chloride and sodium nitrate after the double-displacement reaction.

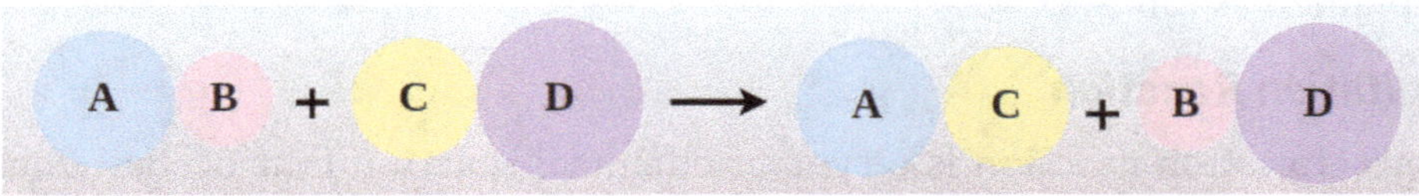

Examples of double displacement reactions are as follows:

- Reaction between silver nitrate and sodium chloride, results in silver chloride and sodium nitrate,

$$AgNO_3\ (aq) + NaCl\ (s) \rightarrow AgCl\ (aq) + NaNO_3\ (s)$$

- Reaction between potassium nitrate and aluminium chloride, results in aluminium nitrate and potassium chloride,

$$KNO_3 + AlCl_3 \rightarrow Al(NO_3)_3 + KCl$$

- Reaction between sodium sulfate and barium chloride, results in barium sulfate and sodium chloride.

$$Na_2SO4\ (aq) + BaCl_2\ (aq) \rightarrow BaSO4\ (s) + 2NaCl(aq)$$

Based on Heat Involved in Reaction

Based on the heat involved in the chemical reaction, Reactions can be classified into two categories:

- Exothermic Reaction
- Endothermic Reaction

Exothermic Reaction

Chemical reactions in which heat energy is released are known as exothermic reactions. The burning of methane gas in the presence of air produces heat energy. All the combustion reactions are exothermic.

For example, the combustion of ethane gives Carbon dioxide and water with some heat as well

$$2C_2H_6\ (s) + 5O_2\ (g) \rightarrow 2CO_2(g) + 6H_2O\ (g) + heat$$

Endothermic Reaction

Chemical reactions that involve the absorption of energy are known as endothermic reactions.

Example: Formation of nitric oxide during the reaction of nitrogen and oxygen. N2 (g) + O_2 (g) + Heat → 2NO

Some Other Chemical Reactions

Other than all the above-mentioned types, there are some more important types of reactions that are used throughout the study of chemistry. These types are explained as follows:

Precipitation Reaction

A precipitation reaction is a type of chemical reaction that occurs when two solutions are mixed and a solid substance (precipitate) is formed as a result of the reaction. In this type of reaction, the ions of two different aqueous solutions react with each other to form an insoluble solid. This solid substance is called a precipitate and it settles at the bottom of the container. An example of a precipitation reaction is the reaction between silver nitrate (AgNO3) and sodium chloride (NaCl):

$$AgNO_3\ (aq) + NaCl\ (aq) \rightarrow AgCl\ (s) + NaNO_3\ (aq)$$

In this reaction, silver nitrate and sodium chloride are mixed together, and the silver ions (Ag+) and chloride ions (Cl-) react with each other to form solid silver chloride (AgCl), which appears as a white precipitate. The sodium and nitrate ions remain in the solution as they do not react with each other.

Neutralization Reaction

A Neutralization reaction is a reaction between an acid and a base to give salt and water as the products. The formation of the water molecules is by the combination of OH– ions and H+ ions. The overall pH of the products when a strong base and a strong acid undergo a neutralization reaction must be 7.

Consider the example of the neutralization reaction between Hydrochloric acid (HCl) and Sodium Hydroxide (NaOH) giving out sodium chloride (Common Salt) and water.

$$HCl\ (l) + NaOH\ (aq) \rightarrow NaCl\ (aq) + H2O\ (l)$$

Combustion Reaction

A combustion reaction is a type of chemical reaction between a fuel and an oxidant that produces a product called ad oxidized product. an oxidizer is a chemical fuel required to burn such as oxygen. Generally, it's a reaction between hydrocarbon and oxygen to yield carbon dioxide, water, and heat.

Hydrocarbon + Oxygen → Carbon dioxide + Water + Heat

For example, consider the combustion of methane gas,

$$CH_4 (g) + 2O_2 (g) \rightarrow CO_2 (g) + 2H_2O (g) + Heat$$

Combustion of Magnesium ribbon,

$$2Mg + O_2 \rightarrow 2MgO + Heat$$

2

THERMODYNAMICS OF CHEMICAL REACTIONS

GIBBS FREE ENERGY AND SPONTANEITY

Introduction

If the value of Gibbs free energy is known the direction and movement of spontaneous change can be known.At constant temperature and pressure, a spontaneous change in a system goes in the direction of decreasing free energy. In general, the total entropy change is the primary criterion for determining the spontaneity of any process. We may claim there is a change in enthalpy along with the change in entropy since most chemical processes fall into the closed system and open system categories.

Entropy defines the randomness of molecules, thus entropy changes and either enhances or lowers the randomness. This randomness of molecules can affect the movement. As only the randomness of a molecule cannot define a process to be spontaneous thus, Gibbs free energy is used to know whether the process is spontaneous or not.

Gibbs free energy, also known as Gibbs function, Gibbs energy, or free enthalpy, is a term used to measure the greatest amount of work done in a thermodynamic system when temperature and pressure remain constant. The sign 'G' stands for Gibbs free energy. Its energy is often measured in Joules or Kilojoules. The maximum amount of work that may be extracted from a closed system is defined as Gibbs free energy. The Gibbs free energy is determined using Gibbs Helmholtz equation defined as G=TH-S. Where T is the temperature, H is the change in enthalpy, S is the change of entropy and G is the change in Gibbs free energy.

Josiah Willard Gibbs, an American scientist, discovered this trait in 1876 while performing experiments to anticipate the behaviour of systems when they are joined or whether a process can occur concurrently and spontaneously. "Available energy" was another name for Gibbs free energy. It can be thought of as the quantity of useful energy available in a thermodynamic system that can be put to work.

Spontaneity

Let's try to grasp the concept of spontaneity. A spontaneous process is irreversible, and it can only be stopped by external forces. The degree of randomness in a system is measured by its entropy.

The Gibbs equation allows us to predict reaction spontaneity directly from enthalpy and entropy measurements. When a reaction is exothermic, the system's enthalpy is negative, resulting in a negative Gibbs free energy. As a result, all exothermic reactions are considered spontaneous. When the enthalpy of the system is positive, endothermic reactions occur spontaneously under two conditions:

The temperature is extremely high, causing the Gibbs energy value to be negative. To make the Gibbs free energy negative, the entropy shift must be quite high.

Spontaneity can only indicate whether or not a reaction can occur, not whether or not it will. At Standard Temperature and Pressure (STP), for example, the spontaneous conversion of diamond to graphite occurs, but it is a gradual process. The makeover will take years to complete.

Entropy

The calculation of a device's thermal energy per unit temperature inaccessible for usable work is known as entropy. Since research is produced from organized molecular motion, the sum of entropy is also a measure of a system's molecular disorder or randomness.

Predicting a reaction's spontaneity

In general, the total entropy change is the most important criterion for determining the spontaneity of any process. We may claim there is a change in enthalpy along with the change in entropy since most chemical processes fall into the closed system and open system categories. Because entropy change affects molecular movements and hence enhances or lowers randomness, entropy change alone cannot account for the spontaneity of such a process. As a result, the Gibbs energy change is used to explain the spontaneity of a process. If the value of $G<0$, then the process is spontaneous and if $G>0$ then the process is non spontaneous. Gibbs' energy is a state function with a wide range of applications.

Spontaneous Reaction

A spontaneous reaction is one in which the production of products is favoured under the conditions of the reaction. A blazing blaze is an example of a spontaneous reaction (see graphic below). A fire is exothermic, which implies that the system's energy decreases when heat is expelled into the environment. Because the majority of a fire's products are gases like carbon dioxide and water vapor, the system's entropy rises throughout most combustion reactions. Combustion processes occur spontaneously due to the combination of a drop in energy and a rise in entropy.

Non-spontaneous reaction

A nonspontaneous reaction is one in which the generation of products is not favoured under the given conditions. One or both of the driving factors must prefer the reactants over the products for a reaction to be nonspontaneous. To put it another way, the reaction is endothermic, has a drop-in entropy, or both.

It's important not to conflate the term spontaneous with the idea that a reaction happens quickly. Even though the reaction is exceedingly slow, product formation is preferred in a spontaneous reaction.

Gibbs Energy

Generally, the total entropy change is the fundamental parameter that defines the spontaneity of any process.

Since most of the chemical reactions fall under the category of a closed system and open system; we can say there is a change in enthalpy too along with the change in entropy.

Gibbs energy is a state function and an extensive property. Since, change in enthalpy too increases or decreases the randomness by affecting the molecular motions, entropy change alone cannot account for the spontaneity of such process. Therefore, we use the Gibbs energy change for explaining the spontaneity of a process.

Expression for Gibbs Energy Change

General expression for Gibbs energy change at constant temperature is expressed as:

$$\Delta G\ sys = \Delta Hsys - T\Delta Ssys$$

Where,

ΔGsys-Gibbs energy change of the system

ΔHsys=enthalpy change of the system

ΔSsys=entropy change of the system

T=Temperature of the system

Above equation is popularly known as Gibbs equation. Gibbs' equation relates enthalpy and entropy of the system. We know that for a spontaneous process, the total entropy change, ΔS is greater than zero.

$\Delta S_{total} = \Delta S_{sys} + \Delta S_{surr}$

Where,

ΔS_{total}-total entropy change for the process

ΔS_{sys}=entropy change of the system

ΔS_{surr}= entropy change of the surrounding

In case of thermal equilibrium between system and surrounding, temperature change between system and surrounding, $\Delta T = 0$. Hence, we can say that enthalpy lost by the system is gained by the surrounding. Hence, the entropy change of the surrounding is given as,

$\Delta S_{surr} = HsurrT = -DHsysT$

$\Delta S_{total} = \Delta S_{sys} + (-\Delta HsysT)$

ΔH_{surr}=change in enthalpy of the surrounding

ΔH_{sys}=change in enthalpy of the system

As discussed earlier, for the spontaneity of a process, $DStotal > 0$. Above equation becomes,

$T\Delta S_{sys} - DH_{sys} > 0$

$\Delta H_{sys} - TDS_{sys} < 0$

The Spontaneity of a Process

Gibbs equation helps us to predict the spontaneity of reaction on the basis of enthalpy and entropy values directly. When the reaction is exothermic, enthalpy of the system is negative making Gibbs free energy negative. Hence, we can say that all exothermic reactions are spontaneous.

In the case of endothermic reactions, when enthalpy of the system is positive, the process is spontaneous under two conditions:

- Temperature is very high to make the Gibbs energy value negative
- Entropy change is very high to make the Gibbs free energy negative.

Spontaneity can only indicate if a reaction can occur not necessarily if a reaction will occur. For example, the conversion of diamond to graphite is a spontaneous process at Standard Temperature and Pressure (STP) but it is a slow process. It will take years for the transformation to occur.

Gibbs Energy in Chemistry

The energetics of processes for systems at constant temperature and pressure, the appropriate quantity is known as the Gibbs free energy. The Gibbs free energy has a very useful property; it decreases for a spontaneous process at constant temperature and pressure. Under such conditions the decrease in Gibbs free energy equals the maximum amount of energy available for work, whereas if it increases for some transition, the change in Gibbs free energy represents the minimum amount of work required.

The transformation of a system from one state to another, at constant temperature and pressure, is spontaneous if the Gibbs free energy decreases. If the Gibbs free energy is unchanged by the transformation the two states are in equilibrium. In other words the criterion for the thermodynamic equilibrium of a system at constant temperature and pressure is that the Gibbs free energy of the system be at the minimum value. The Gibbs free energy is sometimes called the thermodynamic potential at constant pressure in order to indicate its analogy with the potential energy of a mechanical system, which also has a minimum value under equilibrium conditions.

ENTHALPY AND ENTROPY CHANGES IN REACTIONS

Enthalpy and Entropy are important concepts in thermodynamics that relate heat and work. Enthalpy can be defined as the total energy in a system, whereas entropy is defined as the thermal energy of a system per unit temperature. Thus the major difference between enthalpy and entropy is that enthalpy is a type or kind of energy of the system whereas entropy is only a property of the system as a whole.

In this section, we will learn about Enthalpy, Entropy, the Difference between Enthalpy and Entropy, and others in detail.

Enthalpy

Enthalpy is the total amount of energy contained in the system. It can also be defined as the sum of the internal energy of the system and the product of pressure and volume of the system. The heat in a system may enter or leave a system. Thus enthalpy is an important factor in understanding the nature of chemical reaction.

Enthalpy Formula

If we denote enthalpy with H and internal energy, pressure and volume with U, P and V respectively, then the relation between them can be mathematically written as:

$$H = U + PV$$

If we use enthalpy change, internal change and volume change then the same equation can be written as:

$$DH = \Delta U + P\Delta V$$

If enthalpy change is negative, then heat is given out and reaction is exothermic in nature whereas heat is absorbed if enthalpy change is positive and reaction is endothermic reaction.

Examples of Enthalpy

Various real life examples of enthalpy are as follows:

- The number of calories in food is estimated by the amount of energy released due to breaking of glucose bonds in the body.
- Efficiency of engines of automobiles are estimated with the help of enthalpy change.
- Enthalpy is also used to design the refrigerant compressors.

Entropy

Entropy of a system represents the degree of randomness of a system. Higher the randomness of the molecules in a system, more is the entropy of the system. Thus gases have the highest entropy among 3 states of matter. Entropy may also be defined as the thermal energy of a system per unit temperature.

Entropy Formula

If we denote entropy by S, internal energy by Q and temperature by T, then relation between them can be mathematically represented as:

$$S = Q/T$$

If we use entropy change, internal energy change and temperature then the Entropy Change Formula is,

$$DS = \Delta Q/T$$

Note: Entropy of the universe is always increasing.

Examples of Entropy

Some real life examples of entropy are as follows:

- The smoke of an incense cone always expands and spreads in the room which shows the increase in entropy.
- Universe always favors the process which increases entropy. For example, melting of ice is a natural phenomenon and happens on its own as it leads to increase in entropy of the ice when it changes from solid state to liquid state.

Enthalpy and Entropy

Relation between enthalpy and entropy can be mathematically expressed using Gibbs Helmholtz equation which is written as follows:

$$\Delta G = DH - TDS$$

Where,

- ΔG is the change in free energy,
- ΔH is the enthalpy change, and
- ΔS is the entropy change.

Note: DG is always negative for a spontaneous reaction.

Relationship between enthalpy and entropy

Gibbs Free Energy is used to measure the amount of available energy that a chemical reaction provides. As reactions are usually temperature dependent, and sometimes work significantly better at some temperatures than others, the DGf° values known are only valid at 25°C (298.15 K).

Similar to the equations for DH and DS for a system, DG is defined as the difference between the sum of the free energy of formation values of the products and reactants:

$$\Delta G \text{ reaction} = \sum \Delta G \text{ products} - D\sum G \text{ reactants}$$

or simplified

$$DG = \sum DG - D\sum G$$

If a reaction is not spontaneous, its ΔG will be positive. If a reaction is spontaneous, its ΔG will be negative.

It's important to note that spontaneous does not necessarily mean that the specific reaction proceeds at high rate. A spontaneous reaction can take ages to go to completion. A classic example is the rusting of metal.

Going back to enthalpy and entropy, we can define the relationships between these two values, correlating them with the Gibbs free energy. For all temperatures, including 25°C, the following equation can be used to determine spontaneity of a chemical reaction:

$$\Delta G = \Delta H - T\Delta S$$

This equation is valid only if:

- The temperature is in Kelvin, which is done by adding 273.15 to the Celsius temperature.
- S of the reaction is converted to kJ/K.

The value calculated for ΔG is considered an approximate, especially as the temperature moves further away from 25°C as both ΔH and ΔS will vary with temperature. A change of ΔS will impact ΔG tends less. This is because ΔS is measured in units of J/K and when converted to kJ/K it is numerically small. A small change of ΔH but can have a great impact on ΔG.

STANDARD FREE ENERGY CHANGE AND EQUILIBRIUM

Energy can take many forms, including kinetic energy produced by an object's movement, potential energy produced by an object's position, heat energy transferred from one object to another due to a temperature difference, radiant energy associated with sunlight, the electrical energy produced in galvanic cells, the chemical energy stored in chemical substances, and so on. All of these different types of energy may be transformed from one form to the other.

For example, as water in a dam reservoir falls, its potential energy is turned into kinetic energy, and if the falling water is utilized to power turbines, the kinetic energy of the water is converted into electrical energy. However, if the water collides with rocks near the dam's base. The kinetic energy of the object is transformed into thermal energy.

As a result, the various kinds of energy are quantitatively tied to one another. Thermodynamics is the study of such quantitative relationships between various kinds of energy. Energy shifts occur as a result of physical and chemical processes. The study of energy transitions in these processes is the main focus of thermodynamics.

Gibbs Energy

J.W. Gibbs was an American theoretician. He introduced a new thermodynamic function called Gibbs energy denoted as G. The second law of thermodynamics states that $\Delta S_{total} = (\Delta S_{system} + \Delta S_{Surrounding})$ must be positive for all spontaneous processes. To assess the spontaneity of a process, two entropy changes, ΔS_{System} and $\Delta S_{Surrounding}$ must be determined. As a result, it is more straightforward to describe the criteria of spontaneity in terms of the system's thermodynamic features alone, without respect for the surroundings and this problem was solved by Gibbs.

Gibbs energy (G) is defined as,

$$G = H - TS$$

where:

- S is the entropy of the system,
- H is Enthalpy, and
- T is the Temperature.

G is also a state function because, H, T, and S are state functions.

The change in Gibbs Energy (ΔG) on the initial and final state of the system and not on the path connecting the two states. The change in Gibbs energy at constant temperature and pressure is defined as:

$$\Delta G = \Delta H - T\,\Delta S$$

where:

- ΔS is the change in entropy of the system and
- ΔH is the change in Enthalpy.

Gibbs Energy and Spontaneity

The total entropy change can be written as,

$$\Delta S_{total} = \Delta S_{system} + \Delta S_{surrounding} = \Delta S + \Delta S_{surr}$$

According to the second law of thermodynamics, DStotal > 0 at constant temperature and pressure for the process to be spontaneous. If DH is the enthalpy change accompanying the reaction, that is the enthalpy change of the system then the change in enthalpy of the surroundings is ($-\Delta H$)

Therefore,

$$\Delta S_{surr} = -DH / T$$

Hence the total entropy is given by,

$$\Delta S_{total} = DS - DH / T$$

This equation shows that ΔS_{total} is expressed in terms of the properties of the system only.

Rearranging the equation we get,

$$-T\,\Delta S_{total} = -T\,DS + DH$$

or

$$-T\,\Delta S_{total} = DH - T\,\Delta S$$

Combining the above two equations, we get,

$$\Delta G = -T\,\Delta S_{total}$$

This equation indicates that ΔG and ΔS_{total} have opposite signs because T is always positive. Thus, for a spontaneous process carried out at a constant temperature and pressure $\Delta S_{total} > 0$ and hence DG < 0.

Gibbs energy of a system decreases in a spontaneous change that takes place at constant temperature and pressure. On contrary, for a non-spontaneous reaction DStotal and hence DG > 0.

Gibbs energy of a system increases in a non-spontaneous change that takes place at constant temperature and pressure. The end of the spontaneous process is an equilibrium that corresponds to a minimum in G. Hence the change in Gibbs energy is:

- $\Delta G < 0$, the process is spontaneous.
- $\Delta G > 0$, the process is non-spontaneous.
- $\Delta G = 0$, the process is at equilibrium.

Factor affecting the Spontaneity

Consider the equation,

$$\Delta G = \Delta H - T\,\Delta S$$

The value DG determines whether a physical or chemical change will occur spontaneously. The equations ΔH and ΔS correspond to the values of the system alone. The equation states that two elements influence the spontaneity of reactions

- ΔH is the amount of heat transmitted at constant pressure and temperature, and
- ΔS is the rise or reduction in molecular disorder.

The spontaneous process is favoured by a decrease in enthalpy (-ΔH) and increase in entropy (ΔS) On the other hand non-spontaneous reaction is favoured by an increase in enthalpy (+ΔH) and decrease in entropy (-ΔS). The term temperature in the equation is an important component in determining the relative relevance of the enthalpy and entropy contributions to ΔG. If ΔH and ΔS in the equation are both positive or both negative, the sign of ΔG and hence the spontaneity of the reaction, depends on temperature.

Relating Gibbs Free Energy and Equilibrium constant

The concepts of Gibbs free energy and equilibrium constant are fundamental principles within the realm of thermodynamics.

The determination of the ΔG value for a given chemical reaction provides insight into the thermodynamic favorability of the system towards either the reactants or products at the point of equilibrium. It is known that the equilibrium constant, Keq, establishes a relationship between the concentrations of all substances involved in the reaction when it reaches equilibrium. It is reasonable to assume that there exists a correlation between the ΔG° values and Keq. A more sophisticated analysis of thermodynamics results in the derivation of the subsequent equation:

$$\Delta G^\circ = -RT \ln(K_{eq})$$

In this context, R denotes the ideal gas constant with a value of 8.314 J/K•mol. The variable T represents the temperature in Kelvin, while ln(Keq) signifies the natural logarithm of the equilibrium constant.

In instances where Keq is significant, the reaction products are given preference, thereby indicating a negative DG°. In cases where Keq is small, the reactants are considered to be the preferred species. It can be observed that the logarithm of a value below unity in the natural base is negative. Consequently, in cases where the equilibrium constant (Keq) is less than one, the standard change in Gibbs free energy (ΔG°) exhibits a positive sign.

The following table presents a summary of the correlation between ΔG° and Keq.

K_{eq}	$\ln(K_{eq})$	DG°	Description
>1	positive	negative	Products are favored at equilibrium.
1	0	0	Reactants and products are equally favored.
<1	negative	positive	Reactants are favored at equilibrium.

REACTION QUOTIENT AND LE CHATELIER'S PRINCIPLE

Reaction Quotient

The Reaction Quotient, denoted as Q, is a measure that compares the concentrations of products to reactants in a chemical reaction at any point in time. It's particularly useful for determining the direction in which a reaction will proceed to reach equilibrium.

Introduction to Le Chatelier's Principle

Le Chatelier's Principle is a critical concept in chemistry that explains how a chemical system at equilibrium responds to disturbances. It shows that a system will adjust in a way that counteracts the change applied to it, striving to maintain its equilibrium state.

Connecting Reaction Quotient and Le Chatelier's Principle

The Reaction Quotient (Q) and Le Chatelier's Principle are interconnected in predicting the direction of a chemical reaction. By comparing the value of Q to the equilibrium constant (K), one can determine which way the equilibrium will shift in response to a disturbance.

If QK, the system will shift towards the reactants. This direct comparison allows us to use Le Chatelier's Principle to predict how changes in conditions (e.g., pressure, concentration, temperature) will influence the outcome of a reaction.

Example: Imagine a reaction system where Q<K due to an addition of reactants. According to Le Chatelier's Principle, the system will attempt to reduce the impact of this change by shifting the equilibrium towards the products, thus increasing the value of Q until it equals K.

Le Chatelier's Principles

Le Chatelier's principles, also known as the equilibrium law, are used to predict the effect of some changes on a system in chemical equilibrium (such as the change in temperature or pressure). The principle is named after the French chemist Henry Louis Le Chatelier.

When factors like concentration, pressure, temperature, and inert gases that affect equilibrium are changed, the equilibrium will shift in that direction, where the effects caused by these changes are nullified.

Le Chatelier's principles are often used to manipulate reversible reactions in order to obtain suitable outcomes (such as an improvement in yield).

Effect of Concentration Changes on Equilibrium and Product Formation

As per Le Chatelier's principles, the only way for equilibrium to accept more reactants is to increase product formation. The forward reaction is favoured when the concentration of the reactant is increased. The equilibrium of the reaction shift towards the use of reactants in the reaction, which decreases the concentration of the reactants.

Similarly, the addition of product (concentration/pressure) shall increase the backward reaction to decrease the product concentration. The backward reaction is favoured when the concentration of the reactant decreases, the equilibrium of the reaction shift towards the production of reactants, and the concentration of the reactants is more.

Example:

Consider a reaction between oxygen and sulfur dioxide to produce sulfur trioxide. $2SO_2(g) + O_{2(g)}$ Ë! $2SO_{3(g)}$

If the concentration of the reactant increases, then

- Equilibrium will shift towards the decrease in the concentration of the reactants.
- More favoured for the forward reaction.
- Some of the SO_2 or O_2 form SO_3.
- Equilibrium of the reaction shift towards the right.

If the concentration of the reactant decreases, then

- Equilibrium will shift towards the increase in the concentration of the reactants.
- More favoured for the backward reaction.
- Some of the SO_3 would change to SO_2 or O_2.

- Equilibrium of the reaction shift towards the left.

If the product decreases, then

- The equilibrium of the reaction shift to increase the concentration of the sulfur trioxide.
- Increase in the forward reaction rate.
- Some of the SO_2 or O_2 form SO_3.
- Equilibrium of the reaction shift towards the right.

If the product increases, then

- The equilibrium of the reaction shifts to decrease the concentration of the sulfur trioxide.
- Increase in the reverse reaction rate.
- Some of the SO_3 would change to SO_2 or O_2.
- Equilibrium of the reaction shift towards the left.

Effect of Change of Volume, Pressure, or Inert Gas on Equilibrium and Product Formation

$$K_p = K_c (RT)^{\Delta n} = Kc (p/v)^{\Delta n}$$

Change of volume, pressure or inert gases has no effect on the reactions of liquids and solids. They may have an effect in gaseous reactions, and that too, only when the difference in the sum of the number of reactant and product molecules (“n) is not zero.

When $\Delta n = 0$:

As per Le Chatelier's principles, there will be no effect on Equilibrium and Product Formation by changing the volume, pressure or inert gas.

When $\Delta n = +ve$:

An increase in pressure or decrease in volume will decrease the formation of the product. A decrease in pressure or increase in volume shall have the opposite effect of increasing the product formation.

Inert gases do not take part in the reaction and shall increase the volume or pressure only.

- At constant pressure, the addition of inert gas increases the volume, so it increases the product formation.
- At constant volume, inert gas is added, which does not take part in the reaction; the equilibrium remains undisturbed because the addition of inert gas at constant volume does not change the partial pressures or the molar

concentrations of the substance involved in the reaction. The reaction quotient changes only if the added gas is a reactant or product involved in the reaction.

$$PCl_5 f\, PCl_3 + PCl_2$$

In the decomposition of phosphorus pentachloride Dn = +1.

An increase in pressure or decrease in volume decreases the decomposition of PCl_5

- At constant pressure, the addition of inert gas increases the PCl_5 formation.
- At constant volume, the addition of inert gas decreases the PCl_5 formation. When Δn = -ve:

As per Le Chatelier's principle, an increase in pressure or decrease in volume will increase the formation of the product.

- At constant pressure, the addition of inert gas increases the volume, so it decreases the product formation.
- At constant volume, the equilibrium remains undisturbed. It is because adding an inert gas at constant volume does not change the partial pressures or the molar concentrations of the substance involved in the reaction. The reaction quotient changes only if the added gas is a reactant or product involved in the reaction.

$$N_2 + 3H_2 \rightarrow 2NH_3$$

In the formation of ammonia, Dn = -2. The increase of pressure or decrease in volume increases the formation of ammonia.

- At constant pressure, the addition of inert gas decreases ammonia formation.
- At constant volume, the addition of inert gas increases ammonia formation.

Effect of Change of Temperature on Equilibrium and Product Formation

The individual reaction in the equilibrium can be either endothermic or exothermic. Likewise, at equilibrium, the net energy involved may make the reversible reactions either endothermic or exothermic.

According to Le Chatelier's Principles,

- In exothermic equilibrium, an increase in temperature decreases product formation, and a decrease in temperature increases product formation.
- In endothermic reactions, an increase in temperature increases the product formation, and a decrease in temperature decreases the product formation

$$\log\left[\frac{K_1}{K_2}\right] = \frac{-\Delta H^o}{R}\left[\frac{1}{T_2} - \frac{1}{T_1}\right]$$

Le Chatelier's Principle on Change of Temperature

As per the Van't Hoff equation, for an exothermic equilibrium, "H will be negative. An increase in temperature shall decrease K2, or a decrease in temperature increases K2. The opposite is true for an endothermic reaction.

Example:

Consider a reaction

$$N_{2(g)} + 3H_{2(g)} \rightarrow 2NH_{3(g)} \ \Delta H = -92kJ$$

Increase in temperature

- This prefers an endothermic reaction because it takes energy.
- An endothermic reaction is a reverse reaction, and it is favoured.
- The yield of the product (NH_3) decreases.

Decrease in temperature

- This prefers an exothermic reaction because it gives energy.
- An exothermic reaction is a forward reaction, and it is favoured.
- The yield of the product (NH_3) increases.

3

KINETICS OF CHEMICAL REACTIONS

RATE LAWS AND REACTION ORDER

Rate Law

The rate law (also known as the rate equation) for a chemical reaction is an expression that provides a relationship between the rate of the reaction and the concentrations of the reactants participating in it.

Rate laws (sometimes called differential rate laws) or rate equations are mathematical expressions that describe the relationship between the rate of a chemical reaction and the concentration of its reactants. As an example, consider the reaction described by the chemical equation

$$aA+bB \rightarrow products$$

where a and b are stoichiometric coefficients. The rate law for this reaction is written as:

$$rate=k[A]_m[B]_n$$

in which [A] and [B] represent the molar concentrations of reactants, and k is the rate constant, which is specific for a particular reaction at a particular temperature. The exponents m and n are the reaction orders and are typically positive integers, though they can be fractions, negative, or zero. The rate constant k and the reaction orders m and n must be determined experimentally by observing how the rate of a reaction changes as the concentrations of the reactants are changed. The rate constant k is independent of the reactant concentrations, but it does vary with temperature.

The reaction orders in a rate law describe the mathematical dependence of the rate on reactant concentrations. Referring to the generic rate law above, the reaction is m order with respect to A and n order with respect to B. For example, if m = 1 and n = 2, the reaction is first order in A and second order in B. The overall reaction order is simply the sum of orders for each reactant. For the example rate law here, the reaction is third order overall (1 + 2 = 3). A few specific examples are shown below to further illustrate this concept.

The rate law:

$$rate = k[H_2O_2]$$

describes a reaction that is first order in hydrogen peroxide and first order overall. The rate law:

$$rate=k[C_4H_6]_2$$

describes a reaction that is second order in C4H6and second order overall. The rate law:

$$rate=k[H+][OH-]$$

describes a reaction that is first order in H+, first order in OH'', and second order overall.

Example

Writing Rate Laws from Reaction Orders

An experiment shows that the reaction of nitrogen dioxide with carbon monoxide: NO2(g) + CO(g) → NO(g) + CO2(g) is second order in NO2and zero order in CO at 100 °C.What is the rate law for the reaction?

Solution

The reaction will have the form:

$$rate=k[NO2]_m[CO]_n$$

The reaction is second order in NO2; thus m = 2. The reaction is zero order in CO; thus n = 0. The rate law is:

$$rate=k[NO_2]2[CO]0=k[NO_2]$$

Remember that a number raised to the zero power is equal to 1, thus [CO]0 = 1, which is why the CO concentration term may be omitted from the rate law: the rate of reaction is solely dependent on the concentration of NO2. A later chapter section on reaction mechanisms will explain how a reactant's concentration can have no effect on a reaction rate despite being involved in the reaction.

Reaction Order and Rate Constant Units

The Order of reaction refers to the relationship between the rate of a chemical reaction and the concentration of the species taking part in it. In order to obtain the reaction order, the rate expression (or the rate equation) of the reaction in question must be obtained.

Once the rate equation is obtained, the entire composition of the mixture of all the species in the reaction can be understood.

In some of our examples, the reaction orders in the rate law happen to be the same as the coefficients in the chemical equation for the reaction. This is merely a coincidence and very often not the case.

Rate laws may exhibit fractional orders for some reactants, and negative reaction orders are sometimes observed when an increase in the concentration of one reactant causes a decrease in reaction rate. A few examples illustrating these points are provided:

$$NO_2 + CO \rightarrow NO + CO_2 \text{ rate} = k[NO_2]^2$$

$$CH_3CHO \rightarrow CH_4 + CO \text{ rate} = k[CH_2CHO]^2$$

$$2N_2O_5 \rightarrow NO_2 + O_2 \text{rate} = k[N_2O_5]$$

$$2NO_2 + F_2 \rightarrow 2NO_2F \text{rate} = k[NO_2][F_2]$$

$$2NO_2Cl \rightarrow 2NO_2 + Cl_2 \text{rate} = k[NO2Cl]$$

It is important to note that rate laws are determined by experiment only and are not reliably predicted by reaction stoichiometry.

Table 3.1 summarizes the rate constant units for common reaction orders.

Table 3.1 Rate Constant Units for Common Reaction Orders

Overall Reaction Order (x)	Rate Constant Unit (Lx-1 mol1-x s-1)
0 (zero)	mol L^{-1} s^{-1}
1 (first)	s^{-1}
2 (second)	L mol^{-1} s^{-1}
3 (third)	L2 mol^{-2} s^{-1}

Note that the units in this table were derived using specific units for concentration (mol/L) and time (s), though any valid units for these two properties may be used.

How to Determine Reaction Order

There are several different methods which can be followed in order to determine the reaction order. Some of these methods are described in this subsection.

Initial Rates Method

- First, the natural logarithm form of the power-law expression is obtained. It is given by: ln r = ln k + x.ln[A] + y.ln[B] + ...

- The partial order corresponding to each reactant is now calculated by conducting the reaction with varying concentrations of the reactant in question and the concentration of the other reactants kept constant.
- If the partial order of A is being determined, the power-law expression of the rate equation now becomes ln r = x.ln[A] + C, where C is a constant.
- A graph is now plotted by taking 'ln r' as a function of ln[A], the corresponding slope is the partial order, given by x.

Integral Method

- The order of reaction obtained from the initial rates method is usually verified using this method.
- The measured concentrations of the reactants are compared with the integral form of the rate law.
- For example, the rate law for a first-order reaction is verified if the value for ln[A] corresponds to a linear function of time (integrated rate equation of a first order reaction: $\ln[A] = -kt + \ln[A]_0$).

Differential Method

- This method is the easiest way to obtain the order of reaction • First, the rate expression of the reaction is written (r = k[A]x[B]y..)
- The sum of the exponents x+y+... gives the final value of the reaction order.

Apart from these methods, there exist other ways to obtain the reaction order, such as the method of flooding in which the concentration of a single reactant is measured when all the other reactants are present in huge excess.

Different Values of Reaction Order

As discussed earlier, the value of the order of reaction may be in the form of an integer, zero, or a fraction. A graph detailing the reaction rates for different reaction orders can be found below.

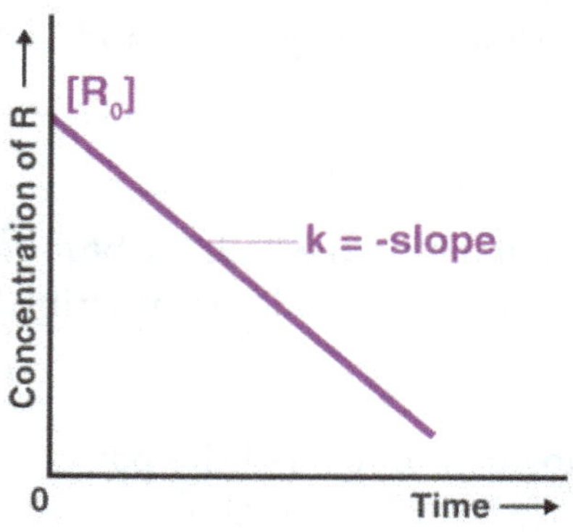

Chemical reactions can be classified into the following types based on the dependence of the rate on the concentration.

Zero Order Reactions

- The rate of reaction is independent of the concentration of the reactants in these reactions.
- A change in the concentration of the reactants has no effect on the speed of the reaction
- Examples of these types of reactions include the enzyme-catalyzed oxidation of CH_3CH_2OH (ethanol) to CH_3CHO (acetaldehyde).

First-Order Reactions

- The rates of these reactions depend on the concentration of only one reactant, i.e. the order of reaction is 1.
- In these reactions, there may be multiple reactants present, but only one reactant will be of first-order concentration while the rest of the reactants would be of zero-order concentration.
- Example of a first-order reaction: $2H_2O_2 \rightarrow 2H_2O + O_2$

Pseudo-First Order Reactions

- In a pseudo-first order reaction, the concentration of one reactant remains constant and is therefore included in the rate constant in the rate expression.
- The concentration of the reactant may be constant because it is present in excess when compared to the concentration of other reactants, or because it is a catalyst.
- Example of a pseudo-first order reaction: $CH_3COOCH_3 + H_2O \rightarrow CH_3COOH + CH_3OH$ (this reaction follows pseudo-first order kinetics because water is present in excess).

Second-Order Reaction

- When the order of a reaction is 2, the reaction is said to be a second-order reaction.
- The rate of these reactions can be obtained either from the concentration of one reactant squared or from the concentration of two separate reactants.
- The rate equation can correspond to $r = k[A]^2$ or $r = k[A][B]$
- Example of a second-order reaction: $NO_2 + CO \rightarrow NO + CO_2$

FACTORS INFLUENCING REACTION RATES

The major factors affecting the collision rate (and hence of a reaction) are the following:

- Concentration of reactants
- Temperature
- Presence of a catalyst
- Spatial orientation of reactant molecules
- Phase of reactants
- Surface area of reactants

Concentration of Reactants

The concentration of reactant molecules has an impact on the reaction rate since the increase in the concentration of particles leads to more collisions between the molecules. The rapid collision of particles makes the reaction occur more often. Therefore, the rate of reaction also increases with the increase of the reactant concentrations.

To better understand the concept, check the image attached below:

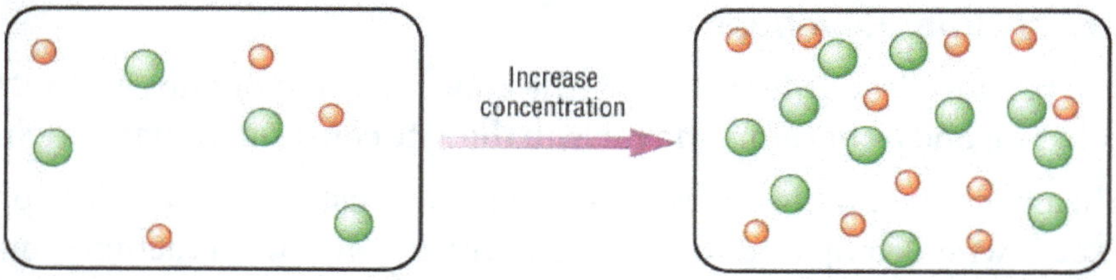

- You can see, have two containers with reactant molecules. In the second container, we have increased the concentration of reactant molecules. It is obvious that there will be more collisions between the reactant molecules in the case of the second container.
- In the case of gaseous samples, the increase in pressure also increases the concentration of the molecules, meaning that the reaction rate is also increased.
- Consequently, we can conclude that if we increase the concentration of reactants, we increase the number of collisions; Thus, the reaction rate increases.

Temperature

As you may already know, reactant molecules need some energy for the reaction to take place. This minimum energy required for the molecules to react with each other is referred to as the activation energy.

The kinetic energy of a gas is directly proportional to temperature; therefore, increasing the temperature also increases the energy which means that the molecules are more likely to collide with each other with sufficient activation energy for the reaction to proceed.

As a result, if we increase the temperature of a system, we can increase the probability of the molecules moving and colliding with the necessary energy.

Let's consider the following diagram (Fig. 3.1) to get a better idea of the relationship between the temperature and activation energy of reactant molecules:

The diagram (Fig. 3.1) represents the number of collisions between the reactant molecules as a function of their corresponding kinetic energies at two different temperature values.

T_1 is the initial temperature of the system, while T2corresponds to the increased temperature. Therefore, T2> T1.

E_a– the activation energy - the minimum energy needed for a reaction to occur. Since the reactant molecules typically collide more rapidly in hot systems, the increase in temperature causes the increase in kinetic energy of the colliding molecules, as it is shown in the diagram above. As a result, the reaction occurs much faster than in the case of cold systems.

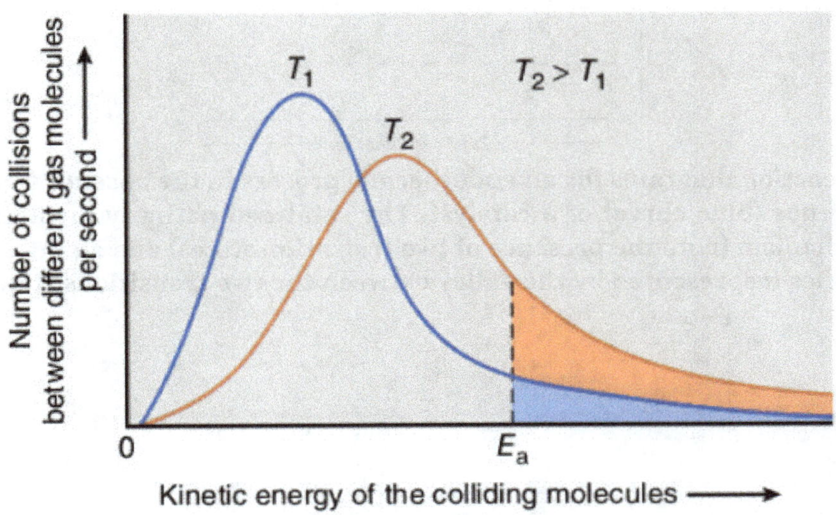

Fig. 3.1 The number of collisions between the reactant molecules as a function of their corresponding kinetic energies at two different temperature values

Catalysts and Inhibitors

Among the factors affecting chemical reaction rates discussed earlier in this chapter was the presence of a catalyst, a substance that can increase the reaction rate without being consumed in the reaction. The concepts introduced in the previous section on reaction mechanisms provide the basis for understanding how catalysts are able to accomplish this very important function.

Figure 3.2 shows reaction diagrams for a chemical process in the absence and presence of a catalyst. Inspection of the diagrams reveals several traits of these reactions. Consistent with the fact that the two diagrams (Fig. 3.3) represent the same overall reaction, both curves begin and end at the same energies (in this case, because products are more energetic than reactants, the reaction is endothermic). The reaction mechanisms, however, are clearly different. The uncatalyzed reaction proceeds via a one-step mechanism (one transition state observed), whereas the catalyzed reaction follows a two-step mechanism (two transition states observed) with a notably lesser activation energy.

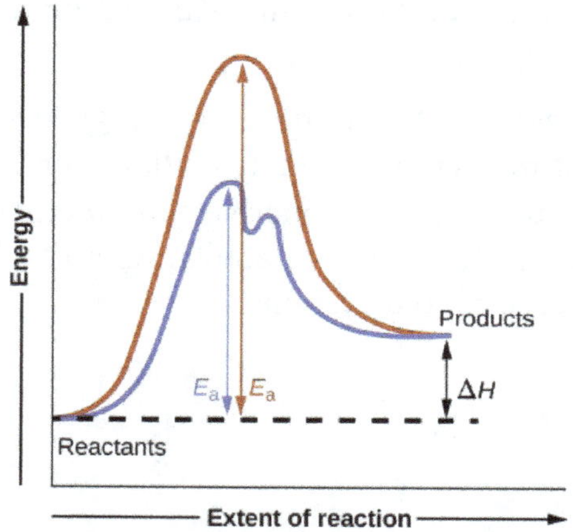

Fig 3.2 Reaction diagrams for an endothermic process in the absence (red curve) and presence (blue curve) of a catalyst. The catalyzed pathway involves a two-step mechanism (note the presence of two transition states) and an intermediate species (represented by the valley between the two transitions states).

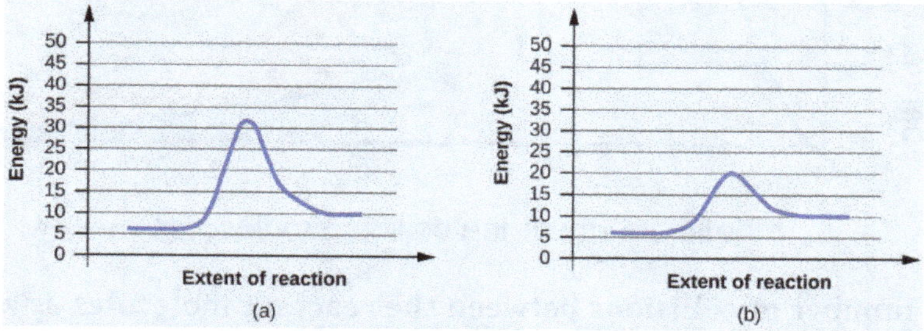

Fig. 3.3 The same reaction: one without a catalyst and one with a catalyst. Estimate the activation energy for each process, and identify which one involves a catalyst.

This difference illustrates the means by which a catalyst functions to accelerate reactions, namely, by providing an alternative reaction mechanism with a lower activation energy. Although the catalyzed reaction mechanism for a reaction needn't necessarily involve a different number of steps than the uncatalyzed mechanism, it must provide a reaction path whose rate determining step is faster (lower Ea).

Reaction Diagrams for Catalyzed Reactions

The two reaction diagrams (Fig. 3.3 above) represent the same reaction: one without a catalyst and one with a catalyst. Estimate the activation energy for each process, and identify which one involves a catalyst.

Homogeneous Catalysts

A homogeneous catalyst is present in the same phase as the reactants. It interacts with a reactant to form an intermediate substance, which then decomposes or reacts with another reactant in one or more steps to regenerate the original catalyst and form product.

As an important illustration of homogeneous catalysis, consider the earth's ozone layer. Ozone in the upper atmosphere, which protects the earth from ultraviolet radiation, is formed when oxygen molecules absorb ultraviolet light and undergo the reaction:

$$3O2(g) \xrightarrow{hv} 2O3(g)$$

Ozone is a relatively unstable molecule that decomposes to yield diatomic oxygen by the reverse of this equation. This decomposition reaction is consistent with the following two-step mechanism:

$$O3 \rightarrow O_2 + O$$

$$O + O_3 \rightarrow 2O_2$$

A number of substances can catalyze the decomposition of ozone. For example, the nitric oxide-catalyzed decomposition of ozone is believed to occur via the following three-step mechanism:

$$NO(g) + O_3(g) \rightarrow NO_2(g) + O_2(g)$$

$$O_3(g) \rightarrow O_2(g) + O(g)$$

$$NO_2(g) + O(g) \rightarrow NO(g) + O_2(g)$$

As required, the overall reaction is the same for both the two-step uncatalyzed mechanism and the three-step NO-catalyzed mechanism:

$$2O_3(g) \rightarrow 3O_2(g))$$

Notice that NO is a reactant in the first step of the mechanism and a product in the last step. This is another characteristic trait of a catalyst: Though it participates in the chemical reaction, it is not consumed by the reaction.

Heterogeneous Catalysts

A heterogeneous catalyst is a catalyst that is present in a different phase (usually a solid) than the reactants. Such catalysts generally function by furnishing an active surface upon which a reaction can occur. Gas and liquid phase reactions catalyzed by heterogeneous catalysts occur on the surface of the catalyst rather than within the gas or liquid phase.

Heterogeneous catalysis typically involves the following processes:

- Adsorption of the reactant(s) onto the surface of the catalyst
- Activation of the adsorbed reactant(s)
- Reaction of the adsorbed reactant(s)
- Desorption of product(s) from the surface of the catalyst

Figure 3.4 illustrates the steps of a mechanism for the reaction of compounds containing a carbon–carbon double bond with hydrogen on a nickel catalyst. Nickel is the catalyst used in the hydrogenation of polyunsaturated fats and oils (which contain several carbon–carbon double bonds) to produce saturated fats and oils (which contain only carbon–carbon single bonds).

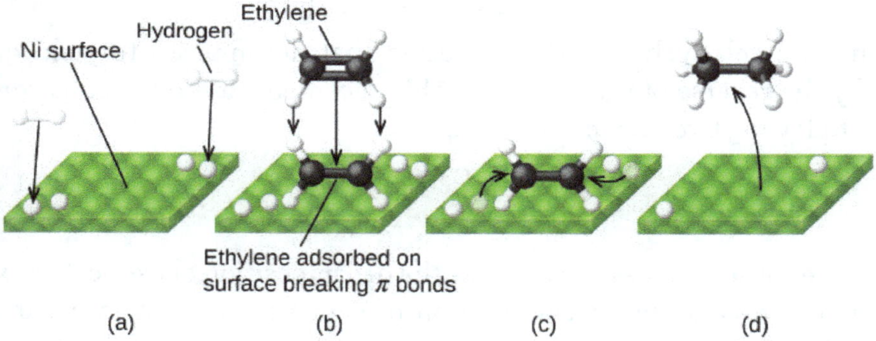

Fig. 3.4 Mechanism for the Ni-catalyzed reaction C2H4+H2→C2H6. (a) Hydrogen is adsorbed on the surface, breaking the H–H bonds and forming Ni–H bonds. (b) Ethylene is adsorbed on the surface, breaking the C–C p-bond and forming Ni–C bonds. (c) Atoms diffuse across the surface and form new C–H bonds when they collide. (d) C2H6 molecules desorb from the Ni surface.

Many important chemical products are prepared via industrial processes that use heterogeneous catalysts, including ammonia, nitric acid, sulfuric acid, and methanol. Heterogeneous catalysts are also used in the catalytic converters found on most gasoline powered automobiles.

Inhibitor

An inhibitor is a chemical compound that can either stop or decrease the reaction rate of a chemical reaction. Therefore, we call them "negative catalysts". Moreover, this compound can reduce the activity of a catalyst as well.

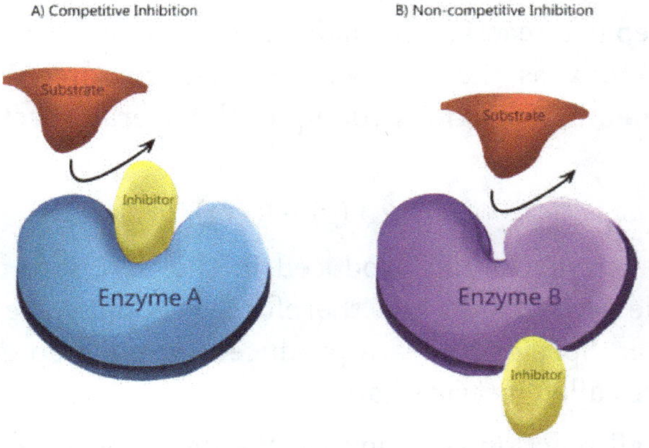

Fig. 3.5: Competitive and Non-Competitive Inhibitors for Enzymes

Unlike catalysts, these compounds do not provide a reaction pathway to reduce the activation energy. The role of an inhibitor is to either deactivate the catalyst or to remove reaction intermediates.

Functioning of Inhibitor

Inhibitors are chemical compounds that slow down the reaction. They sometimes even stop the reaction. They work exactly opposite of catalysts.

Difference Between Catalyst and Inhibitor

Catalyst is a chemical compound that can increase the rate of a reaction without being consumed while Inhibitor is a chemical compound that can either stop or decrease the reaction rate of a chemical reaction. So, this is the key difference between catalyst and inhibitor.

Moreover, catalysts work by providing an alternative reaction pathway by producing a reaction intermediate while inhibitors work by either deactivates the catalyst or removes the reaction intermediates.

REACTION MECHANISMS AND ELEMENTARY STEPS

Chemical reactions very often occur in a step-wise fashion, involving two or more distinct reactions taking place in sequence. A balanced equation indicates what is reacting and what is produced, but it reveals no details about how the reaction actually takes place. The reaction mechanism (or reaction path) provides details regarding the precise, step-by-step process by which a reaction occurs.

The decomposition of ozone, for example, appears to follow a mechanism with two steps:

$$O_3(g) \rightarrow O_2(g) + O$$

$$O + O_3(g) \rightarrow 2O_2(g)$$

We call each step in a reaction mechanism an elementary reaction. Elementary reactions occur exactly as they are written and cannot be broken down into simpler steps. Elementary reactions add up to the overall reaction, which, for the decomposition, is:

$$2O_3(g) \rightarrow 3O_2(g)$$

Notice that the oxygen atom produced in the first step of this mechanism is consumed in the second step and therefore does not appear as a product in the overall reaction. Species that are produced in one step and consumed in a subsequent step are called intermediate.

While the overall reaction equation for the decomposition of ozone indicates that two molecules of ozone react to give three molecules of oxygen, the mechanism of the reaction does not involve the direct collision and reaction of two ozone molecules. Instead, one O3 decomposes to yield O2 and an oxygen atom, and a second O3 molecule subsequently reacts with the oxygen atom to yield two additional O2 molecules.

Unlike balanced equations representing an overall reaction, the equations for elementary reactions are explicit representations of the chemical change taking place. The reactant(s) in an elementary reaction's equation undergo only the bond-breaking and/or making events depicted to yield the product(s). For this reason, the rate law for an elementary reaction may be derived directly from the balanced chemical equation describing the reaction. This is not the case for typical chemical reactions, for which rate laws may be reliably determined only via experimentation.

Unimolecular Elementary Reactions

The molecularity of an elementary reaction is the number of reactant species (atoms, molecules, or ions). For example, a unimolecular reaction involves the reaction of a single reactant to produce one or more molecules of product:

$A \rightarrow$ product

The rate law for a unimolecular reaction is first order; rate = k [A].

A unimolecular reaction may be one of several elementary reactions in a complex reaction mechanism. For instance, the reaction ($O_3(g) \rightarrow O_2(g) + O$) illustrates a unimolecular elementary reaction occurring as a part of a two-step reaction mechanism. However, some unimolecular reactions may be the only step of a single-step reaction mechanism. (In other words, an "overall" reaction may also be an elementary reaction in some cases.) For example, the gas-phase decomposition of cyclobutane, C_4H_8, to ethylene, C_2H_4, is represented by the chemical equation:

$$C4H8(g) \rightarrow 2C2H4(g)$$

This equation represents the overall reaction, describing a unimolecular elementary process. The rate law predicted from this equation, assuming it is an elementary reaction, turns out to be the same as the rate law derived experimentally for the overall reaction, showing first-order behavior:

$$rate = - D[C4H8]/Dt = k[[C4H8]$$

This agreement between observed and predicted rate laws indicates that the proposed unimolecular, single-step process is a reasonable mechanism for the butadiene reaction.

Bimolecular Elementary Reactions

A bimolecular reaction involves two reactant species. For example: A + B → products

In the first type, where the two reactant molecules are different, the rate law is first order in A and first order in B (second-order overall)

$$A + B \rightarrow products$$

$$2A \rightarrow products$$

In the second type, in which two identical molecules collide and react, the rate law is second order in A:

$$rate = k[A][A] = k[A]^2$$

Some chemical reactions occur by mechanisms that consist of a single bimolecular elementary reaction. One example is the reaction of nitrogen dioxide with carbon monoxide:

$$NO_2(g) + CO(g) \rightarrow NO(g) + CO_2(g)$$

Bimolecular elementary reactions may also be involved as steps in a multistep reaction mechanism. The reaction of atomic oxygen with ozone is the second step of a two-step ozone decomposition mechanism:

$$O(g) + O_3(g) \rightarrow 2O_2(g)$$

Termolecular Elementary Reactions

An elementary termolecular reaction involves the simultaneous collision of three atoms, molecules, or ions. Termolecular elementary reactions are uncommon because the probability of three particles colliding simultaneously is very rare.

There are, however, a few established termolecular elementary reactions. The reaction of nitric oxide with oxygen appears to involve termolecular steps:

$$2NO_2(g) + O_2(g) \rightarrow 2NO_2(g)$$

$$rate = k[NO]^2[O_2]$$

Likewise, the reaction of nitric oxide with chlorine appears to involve termolecular steps:

$$2NO(g) + CL_2(g) \rightarrow 2NOCL_2(g)$$

$$rate = k[NO]^2[CL_2]$$

Often one of the elementary steps in a multistep reaction mechanism is significantly slower than the others. Because a reaction cannot proceed faster than its slowest step, this step will limit the rate at which the overall reaction occurs. The slowest elementary step is therefore called the rate-limiting step (or rate-determining step) of the reaction.

4

REACTION MECHANISMS AND PATHWAYS

Reaction mechanism, or reaction pathway, describes the successive steps at the molecular level that take place in a chemical reaction. In each step, molecular bonds are either created or broken. Reaction mechanisms are postulated, and then either supported or disproved but the overarching conditions for defining a reaction mechanism/ reaction pathway is that all of the discrete steps together make sense with regard to the overall reaction equation, and that the rate law for the slowest step complies with the empirically measured overall rate.

The individual steps that make up most reaction mechanisms are described as unimolecular or bimolecular. A unimolecular mechanism occurs when a molecule either dissociates to form two species or undergoes internal rearrangement. The rate equation for a unimolecular reaction is described as first order since it only depends on the concentration of one reactant and is written as

$$d[A]/dt= - k[A]$$

where k is the rate constant. A bimolecular reaction occurs when two molecules combine and the rate equation is

$$d[A]/dt = - k[A][B]$$

(second order) or second order in one of the molecular species,

$$d[A]/dt = - k[A]^2$$

Examples of Reaction Mechanisms and Reaction Pathway

There are a number of mechanisms frequently used to explain a reaction. For example, in nucleophilic substitution reactions, the nucleophile brings an electron pair to the substrate molecule and the leaving group departs with an electron pair. In the Sn2 mechanism, the bimolecular substitution is a single step process in which no intermediate forms. The nucleophile attacks and the leaving group departs simultaneously. In the unimolecular substitution Sn1 mechanism, there is a slow ionization of the substrate, forming a transition state carbonium ion that is quickly attacked by the nucleophile. In this reaction, the rate determining step is the formation of the transition state carbonium ion.

Another example of a common reaction mechanism is electrophilic substitution. As in the Sn2 reaction, in SE2 mechanism, the new bond forms at the same time that the old bond breaks. In the Sn2 mechanism, the incoming group attacks at a position that is 180 degrees from the leaving group, due to repelling electron clouds so the location of the ligands on the molecule is different. In the SE2 mechanism, the incoming group can approach from the same side, leading to retention of configuration.

THE IMPORTANCE OF REACTION MECHANISMS

Fundamental Reaction Understanding

Chemical reactions can be complex and competing reactions are commonplace, further complicating outcomes. With this much complexity and diversity, the chemist must understand the fundamentals of why molecules react, thus enabling logic and predictability in comprehending how a reaction is likely to proceed. By correctly understanding the mechanisms occurring in the elementary reaction steps, one can begin to understand how to affect the overall result.

For example, some elementary steps may result in important transient species or intermediates that are transition states to the final product or side products. Understanding the mechanisms in the sequential elementary steps of a reaction provides insight into how modifying reaction conditions can affect the overall reaction outcome, purity, and reaction rates.

UNIMOLECULAR AND BIMOLECULAR REACTIONS

Molecularity of a Reaction

The number of reacting species (atoms, ions or molecules) taking part in an elementary reaction, which must collide simultaneously in order to bring about a chemical reaction is called the molecularity of a reaction.

Generally for a reaction $aA + bB \rightarrow cC + dD$

Molecularity = $a + b$

Unimolecular Reactions

In a unimolecular reaction, a single reactant molecule rearranges to form one or more product molecules. The rate equation for a unimolecular reaction is rate

= k(A). Examples of unimolecular reactions include radioactive decay, cis-trans isomerization, thermal decomposition, ring opening, and racemization.

The reactions with molecularity 1 are called Unimolecular Reactions examples of which are,

- $NH_4NO_2 \rightarrow N_2 + 2H_2O$
- $Br_2 \rightarrow 2Br$

Bimolecular Reaction

In a bimolecular reaction, two reactant molecules collide and exchange energy, atoms, or groups of atoms. The two reactant molecules could be the same molecule (A + A to P) or different (A + B to P). The rate law for the first type of bimolecular reaction is rate = k(A)(B), and the rate law for the second type is rate = k(A)(B) = k. Bimolecular reactions are common in organic reactions such as nucleophilic substitution.

The reactions with molecularity 2 are called Bimolecular Reactions examples of which are,

$$2HI \rightarrow H_2 + I_2$$

Trimolecular Reaction: The reactions with molecularity 1 are called Trimolecular Reactions examples of which are,

$$2NO + O_2 \rightarrow 2NO_2$$

The probability of three particles colliding and reacting simultaneously and reacting is very low. Therefore a molecularity higher than three is generally not observed.

An elementary reaction is defined as a reaction that occurs in a single step. Many reactions that follow single rate law actually take place in a series of steps. Such reactions are called complex reactions. Each step of a complex reaction is an elementary reaction.

MULTISTEP REACTIONS AND REACTION INTERMEDIATES

Introduction: Multistep reactions

In chemistry, multistep reactions are sequences of chemical reactions whereby one or more intermediates are formed and subsequently consumed. These reactions are characterized by their specific elementary steps, each representing a single molecular event. Elementary steps can combine to form reaction mechanisms, which describe the detailed process through which reactants convert to products.

- Elementary steps may include bond formations or bond breakages.

- Intermediates are usually very reactive molecules, such as free radicals, carbocations, or anions.
- Reaction mechanisms can be unimolecular, bimolecular, or termolecular, depending on the number of molecules involved in the elementary steps.
- The benefits of mastering multistep reactions include:
- Successfully synthesizing complex molecules that are not accessible through single step reactions.
- Enhancing the efficiency of syntheses by minimizing waste and side products.
- Improving the conservation of energy and raw materials, thus promoting greener chemistry.
- Allowing the discovery of novel reactions, leading to breakthroughs in chemical research.

One of the most important applications of kinetics is to the study of reaction mechanisms, or the sequences of steps by which chemical reactions occur. For example, consider the reaction of nitrogen dioxide with carbon monoxide:

$$NO_2(g) + CO(g) \rightarrow NO(g) + CO_2(g)$$

Based on the balanced equation, we might hypothesize that this reaction occurs by a single collision between a molecule of nitrogen dioxide and a molecule of carbon monoxide. In other words, we might hypothesize that this reaction is an elementary reaction.

The Process of a Multistep Reaction Mechanism

Identifying the Intermediates of Multistep Reactions

Successful identification of intermediates in a multistep reaction mechanism is a critical step in understanding the reaction as a whole. These species are formed transiently and are often highly reactive. Identifying them typically involves techniques such as spectroscopy, trapping experiments, or kinetic studies. Being able to detect an intermediate can validate a proposed mechanism and inform modifications to enhance the desirable traits of a reaction.

Common techniques for identifying intermediates include:

- Mass spectrometry, which can provide molecular weight and structural information.
- Nuclear Magnetic Resonance (NMR) spectroscopy, which can give insight into the molecular structure and environment.

- Infrared (IR) spectroscopy, which identifies characteristic vibrational modes of functional groups.
- Trapping experiments, where a reactive intermediate is 'caught' or stabilized for observation.

Visualising Reaction Pathways in Multistep Reactions

Visualising the reaction pathways of multistep reactions is essential to gain an in depth understanding of how a reaction proceeds. This typically involves constructing a reaction coordinate diagram, which graphs the potential energy of the system against the progress of the reaction. Each peak and trough on the diagram corresponds to a transition state or an intermediate, respectively. These diagrams help chemists to predict the rate-determining step and the overall activation energy of the reaction.

Key elements to visualise in a reaction pathway are described in the table below:

Element	Role in a Reaction Pathway
Transition State	High-energy configuration preceding an intermediate
Intermediate	Species existing between two transition states
Activation Energy	Energy required to reach the transition state from reactants
Reaction Coordinate	Represents progress from reactants to products

Computer simulations and computational chemistry tools also play a crucial role in visualising and optimising reaction mechanisms. Techniques like computational modelling can predict reaction outcomes and assist in designing experiments for complex syntheses.

Exploring Examples of Multistep Reactions

Delving into the realm of chemical transformations, multistep reactions serve as the cornerstone for constructing a diverse array of molecules. These reactions, often encountered in the synthesis of organic compounds and pharmaceuticals, provide a canvas where complexity and creativity merge. By threading together a series of individual steps, each with its unique transition states and intermediates, chemists can guide reactants through a choreographed dance to form the desired products with high specificity.

Common Examples of Multistep Reaction in Organic Chemistry

Organic chemistry is replete with examples of multistep reactions that exemplify the ingenuity of chemical synthesis. From the formation of simple hydrocarbons to the assembly of intricate natural products, multistep processes are fundamental to achieving molecular complexity. These reactions often involve a variety of mechanisms, such as nucleophilic substitutions, eliminations, additions, radical reactions and rearrangements. For example, the synthesis of a complex molecule like a steroid or an alkaloid may require numerous reaction steps, protecting group strategies, and careful control of reaction conditions to navigate the path from simple starting materials to complex architectures.

Notable multistep reactions in organic chemistry include:

- The Diels-Alder reaction, which constructs cyclic compounds.
- The Wittig reaction, instrumental in forming carbon-carbon double bonds.
- The Friedel-Crafts acylation, used to introduce acyl groups to aromatic rings.
- The Grignard reaction for forming carbon-carbon bonds.

Many of these reactions are named after the chemists who discovered them, such as the Aldol condensation, a pivotal reaction in the construction of carbon frameworks having both alcohol and aldehyde functional groups. These reactions form the building blocks for more intricate multistep synthesis, enabling the progression towards desired targets with the precision of a maestro conducting an orchestra.

The Role of Transition States in Multistep Reactions

At the heart of every chemical reaction lies the concept of the transition state, a critical and fleeting moment where reactants are converted to products. In multistep reactions, understanding the role of transition states is essential to deciphering reaction pathways and optimising reaction conditions. A transition state corresponds to the highest energy point along the reaction coordinate, acting as a gatekeeper between reactants and products, and therefore is never isolated.

The characteristics of transition states include:

- They have partially formed and broken bonds.
- They represent the energy peak in a reaction energy profile.
- They can be predicted and modelled using quantum chemical calculations.

Exploring the nature of transition states allows chemists to determine the kinetic and thermodynamic feasibility of reactions, influencing which reaction paths may be favoured. The stability and structure of a transition state can be affected by various factors, such as solvent effects, temperature, and catalysts. By tweaking these parameters, chemists can often steer reactions towards more favourable outcomes. Transition state theory also gives rise to the concept of activation energy, defined as the energy difference between the reactants and the transition state. This energy barrier must be overcome for a reaction to proceed, highlighting the crucial role that transition states play in the overall reaction kinetics.

Multistep Reaction Synthesis in Pharmaceutical Chemistry

Pharmaceutical chemistry relies on multistep syntheses to produce therapeutically active compounds that meet stringent efficacy, purity, and safety standards. The journey from an initial drug lead to a marketable medication can be long and fraught with synthetic challenges. Designing a multistep synthesis in this context demands an intimate understanding of reaction mechanisms and the ability to predict the reactivity and selectivity at each stage of the process.

Multistep syntheses in pharmaceutical chemistry may incorporate various reactions to introduce chirality, protect reactive functional groups, and carefully build up the drug molecule with specific stereochemistry. The end goal is to construct molecules with a high degree of structural complexity, possessing the necessary pharmacological properties to interact with biological targets. For instance, the synthesis of complex drugs such as Taxol, an anticancer agent, showcases the intricate dance of multistep reactions to assemble a molecule with multiple stereocentres and rings from simpler substances.

The synthesis sequence must consider:

- Yield of each step and overall process efficiency.
- Purity of the intermediates and final product.
- Scalability for industrial production.
- Cost-effectiveness and environmental impact.
- Regulatory requirements for good manufacturing practices.

How Catalysts Influence Reaction Rates in Multistep Syntheses

In the world of multistep syntheses, catalysts are the unsung heroes that can propel a reaction forward, reduce energy barriers, or change the pathway completely. Catalysts function by providing an alternative reaction route with a lower activation energy, thus increasing the reaction rate without being consumed in the process. Their role in pharmaceutical chemistry is indispensable, as they can enhance the selectivity, efficiency, and sustainability of reactions.

Common types of catalysts used in multistep reactions include:

- Metals and metal complexes, which can facilitate a variety of transformations, including hydrogenations and cross-coupling reactions.
- Organocatalysts, made up of non-metallic organic compounds that are capable of enantioselective catalysis.
- Enzymes, which are biological catalysts that offer extreme specificity and can operate under mild conditions.

- Acid or base catalysts that can protonate or deprotonate reactants, making them more reactive.

Incorporating catalysts in a synthesis can also help to avoid harsh conditions and minimize the formation of by-products, making the process more 'green'. For example, palladium-catalyzed cross-coupling reactions have revolutionized the formation of carbon carbon bonds, and this innovation was recognized with a Nobel Prize in Chemistry. Catalysis not only accelerates reactions but also opens up new possibilities for chemists to build complex molecular scaffolds that might otherwise be inaccessible.

Activation Energy in Multistep Reactions

Activation energy is a core concept in the context of multistep reactions, often acting as a gatekeeper determining the speed and feasibility of chemical processes. In multistep reactions, it is not just a single value but a series of energy barriers that must be surmounted as reactants transition through various intermediates to products. These energy thresholds are pivotal in understanding how reactions proceed and can be manipulated in order to optimise yield and selectivity.

Energy Profiles of Multistep Reactions

Energy profiles of multistep reactions provide a visual representation of the energy changes as reactants transform into products through a series of intermediates. Each step in the reaction is associated with an activation energy barrier, a minimum amount of energy needed to form the transition state prior to yielding the intermediates or products. Chemists use these profiles to deduce which steps are rate-limiting, potentially reversible, or in need of catalytic assistance.

To construct an energy profile, consider the following key points:

- Each intermediate in the reaction mechanism has a corresponding valley in the energy profile.
- Each transition state is represented by a peak between two valleys.
- The highest peak represents the rate-limiting step, with the largest activation energy requirement.
- The difference in energy between reactants and products gives the overall reaction energy change.

Frequently, the energy profile is depicted as a reaction coordinate diagram plotting the potential energy of the species involved against the reaction progress. This diagram illustrates the step-wise nature of the reaction, highlighting that each elementary step has its unique activation energy. Through computational methods and experimental kinetics, chemists can estimate these energy barriers and use them to predict the conditions that might affect the reaction rate, such as temperature or the presence of a catalyst. Real-time analytics, such as in situ spectroscopy, can now provide even more nuanced insights into these profiles.

Reaction Intermediates

Reactive intermediates, in chemistry, are highly reactive, high energy and short-lived molecules that will quickly turn into stable molecules when they are generated in a chemical reaction. In certain cases, they are separated and stored. For example, matrix isolation and low temperatures.

Matrix isolation is a technique that is used experimentally in physics and chemistry that includes a material that has been trapped within an unreactive material. The host matrix generally comprises guest particles that are generally embedded. Guest particles can be molecules, atoms and ions. The guest is isolated within the host matrix.

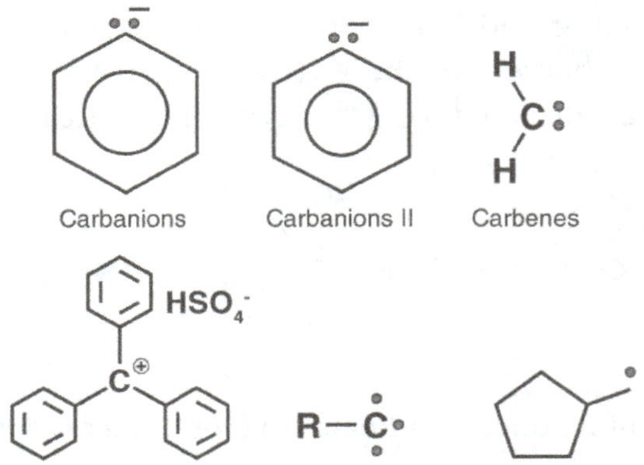

Fig. 4.1 Reactive intermediates

Reactive intermediates occur only in one of the intermediate steps, but in the case of other chemical reactions, they take more than one elementary step. It differs from a simple reaction intermediate or product or reactant only through fast spectrographic methods. Spectroscopy is a study of the interaction between electromagnetic radiation and matter. They are categorised based on the type of radiative energy used in the interaction.

Features of Reactive Intermediates

Its existence can be proved with the help of chemical trapping. The chemical trap is nothing but a chemical compound that detects a molecule in certain cases, and they are listed below.

- When the molecule is highly reactive or when it cannot be determined by spectroscopic means.

- When the concentration of the molecule is below the detection limit.
- When a molecule is present within the mixture, where elements of a component interfere with its detection.

It is hard to distinguish from a transition state. It is a state that corresponds to potential energy at a higher level, along with a reaction coordinate. It is a type of chemical reaction that comprises a particular configuration along with a reaction coordinate. Here, cage effects cannot be taken into consideration. In chemistry, the cage effect illustrates how molecular properties are affected by their surroundings. The cage particle must diffuse from its solvent cage in order to interact with other molecules.

Concerted Reactions

Concerted reactions are chemical reactions that involve only one step. That means; all single-step reactions fall under the category of concerted reactions. Therefore, bond breaking and bond-forming reactions occur simultaneously to form the products. In other words, there are no reactive intermediates or unstable high energy complexes formed during this type of chemical reactions.

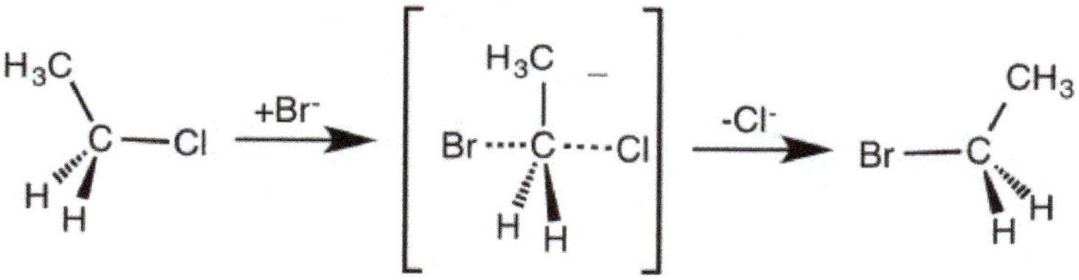

Fig. 4.2: An Example of a Concerted Reaction (contains an unstable intermediate)

Generally, concerted reactions do not depend on the solvent polarity, and the mechanism of the reaction is named as a concerted mechanism. Some common examples for this type of reactions include pericyclic reactions, SN2 reactions and some rearrangement reactions such as Claisen rearrangement.

What are Stepwise Reactions?

Stepwise reactions are multiple-step reactions. Therefore, there are on or more chemical intermediates in these reactions. These intermediates are usually highly reactive due to high energy which makes them unstable. Stepwise reactions typically contain two or more elementary reactions.

In contrast to stepwise reactions, the bond breaking and bond-forming in the stepwise reactions occur in different steps (not in a single step). Therefore, the reactants cannot convert into products directly. The "overall reaction" in a stepwise reaction that gives the conversion of reactants into products. We can get the overall reaction by accurately balancing all the steps in the stepwise reaction. Nevertheless, there is one elementary reaction in a stepwise reaction that determines the rate of the reaction. It is the slowest reaction of the reaction series, and it may or may not have reactants or products involved (it may contain only intermediates of the reaction).

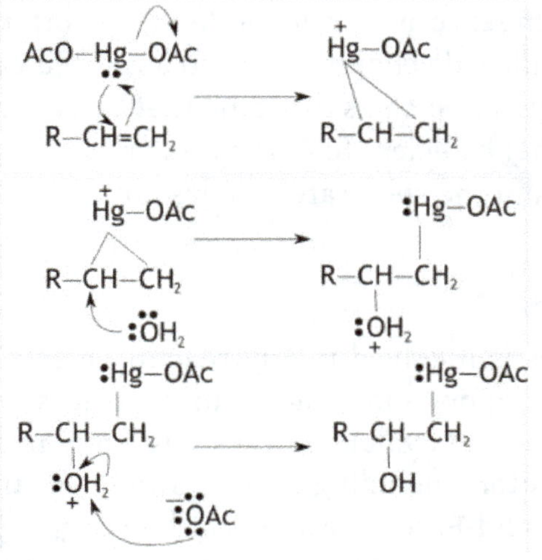

Fig. 4.3: An Example of a Stepwise Reaction

What is the Difference Between Concerted and Stepwise Reactions?

The terms concerted and stepwise reactions come under the field of physical chemistry where the rates of reactions are determined using the change in the reactant and product amounts with time. The key difference between concerted and stepwise reactions is that concerted reactions are single-step reactions, whereas stepwise reactions are multistep reactions. The rate-determining step of a concerted reaction is itself a concerted reaction. But, in stepwise reactions, the slowest elementary reaction is the rate-determining step.

Moreover, there are no reaction intermediates in concerted reactions because the reactants directly convert into the products. However, in stepwise reactions, there are one or more intermediate compounds which are highly unstable due to the high energy level of the compound. The overall reaction of a stepwise reaction gives the conversion of the reactants into products.

CATALYSIS AND ENZYMATIC REACTIONS

Catalytic Mechanisms

The descriptions below describe the major mechanisms enzymes use to catalyze chemical reactions. It should be noted that many enzymes use more than one, and sometimes several different catalytic strategies during their reaction mechanism.

Covalent Catalysis

Covalent catalysis involves the formation of a covalent bond between the enzyme and at least one of the substrates involved in the reaction. Often times this involves nucleophilic catalysis which is a subclass of covalent catalysis. Several amino acid R groups can serve as a nucleophile and are often found at the active site of enzymes. Nucleophilic side chains are often activated by deprotonation caused by neighboring side chains, such as histidine that can act as a base. Alternatively, water can also activate the nucleophile. The intermediate covalent bond formation between the enzyme and the substrate enables bond cleavage and the removal of a leaving group.

Acid-Base Catalysis

Acid-Base Catalysis is involved in any reaction mechanism that requires the transfer of a proton from one molecule to another. It is very common to see this mechanism combined with Covalent Catalysis as many nucleophiles are activated by the removal of a proton, including alcohol, thiol, and amine functional groups. Enzymes that utilize Acid-Base Catalysis can be subgrouped further into either specific acid-base or general acid-base reactions. Specific acid or specific base catalysis occur if a hydronium ion (H3O+) or a hydroxide ion (OH–), respectively, are utilized directly in the reaction mechanism, and the pH of the solution affects the rate of catalysis. General acid and general base reactions occur when molecules other than hydronium ion (H3O+) or a hydroxide ion (OH–) are the source of proton donation or acceptance. Most commonly, an active site amino acid residue is used to accept or donate a proton within the reaction mechanism. In general acid-base reactions the pH is usually held constant within a buffered system.

Electrostatic Catalysis

Electrostatic catalysis occurs when the enzyme active site stabilizes the transition state of the reaction by forming electrostatic interactions with the substrate. The electrostatic interactions can be ionic, ionic-dipole, dipole-dipole, or hydrophobic interactions. Hydrogen bonding is one of the most common electrostatic interactions formed in the active site.

Desolvation

Enzyme active sites can become devoid of water and mimic the reaction characteristics of the gas phase. This can destabilize the polarized state of charged groups such as acids and bases. Thus, the neutral form of these types of residues becomes the favored state. This is due to significant alterations in the pKa of the active site residues within the nonpolar environment. This can cause normally acidic residues such as glutamate to abstract a proton from histidine and behave as a base, for example.

Catalysis by Approximation

In catalysis by approximation, the enzyme enhances the reaction rate by binding with multiple substrates and positioning them favorably so that the reaction can proceed. Binding with the enzyme reduces the rotational entropy of the substrates that would otherwise be randomly free floating in solution, and enables the correct positioning of substrates for the reaction. The loss of entropy, which is not favorable, is offset by the binding energy released with the substrate-enzyme interaction.

In addition to correctly positioning the substrates to interact with one another, catalysis by approximation converts a reaction that would have been second order, with substrates that are free floating in solution, to a first order reaction, where all of the substrates are held in place by the enzyme and behave as a single molecule. This can dramatically improve the catalytic rate of the reaction from 105 to 107 times faster, depending on the enzyme system.

Strain Distortion

In organic chemistry, you learned that certain structures such as three-membered and four-membered ring structures, such as epoxides were highly reactive due to the strain distortion inherent to the unfavored bond angles inherent to the ring. Enzyme active sites can also utilize strain distortion within a bound substrate to increase the reactivity of the molecule and favor the formation of the transition state. Many enzymes that function by the induced fit model also utilize strain distortion within their catalytic mechanism. Within the unbound state they remain in a low catalytic state, however the interaction with the substrate induces the destabilization of the enzyme active site or may induce strain within the substrate causing the initiation of the catalytic activity of the enzyme.

Cofactor Catalysis

Cofactors are molecules that bind to enzymes and are required for the catalytic activity of the enzyme. They can be divided into two major categories: metals and coenzymes. Metal cofactors that are commonly found in human enzymes include: iron, magnesium, manganese, cobalt, copper, zinc, and molybdenum. Coenzymes are small organic molecules that are often derived from vitamins, which are essential organic nutrients consumed within the diet.

Coenzymes can bind loosely with the enzyme and have the ability to bind and release from the active site, or they may be tight binding and lack the ability to release easily from the enzyme. Tight binding coenzymes are referred to as prosthetic groups. Enzymes that are not yet associated with a required cofactor are called apoenzymes, whereas enzymes that are bound with their required cofactors are called holoenzymes. Sometimes organic molecules and metals combine to form coenzymes, such as in the case of the heme cofactor (Fig. 4.4).

Coordination of heme cofactors with their enzyme counterparts often involves electrostatic interactions with histidine residues as shown in the succinate dehydrogenase enzyme shown in Figure 4.4.

Examples of Reaction Mechanisms

Protease Enzymes

Proteolytic enzymes (also termed peptidases, proteases and proteinases) are capable of hydrolyzing peptide bonds in proteins. They can be found in all living organisms, from viruses to animals and humans. Proteolytic enzymes have great medical and pharmaceutical importance due to their key role in biological processes and in the life cycle of many pathogens. Proteases are extensively applied enzymes in several sectors of industry and biotechnology, furthermore, numerous research applications require their use, including production of Klenow fragments, peptide synthesis, digestion of unwanted proteins during nucleic acid purification, cell culturing and tissue dissociation, preparation of recombinant antibody fragments for research, diagnostics and therapy, and the exploration of the structure-function relationships.

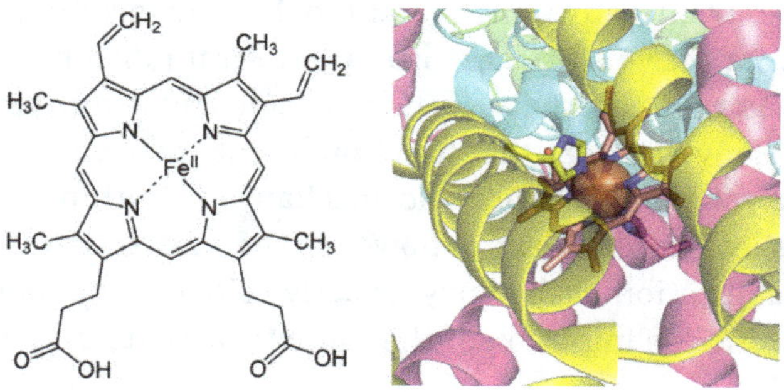

Fig. 4.4 The Heme Cofactor. The family of heme cofactors contain an iron metal coordinated with a porphyrin ring structure as shown in the left hand panel within the structure of Heme B. In the right hand panel, Heme B is shown complexed with the succinate dehydrogenase enzyme from the Kreb Cycle.

Proteolytic enzymes belong to the hydrolase class of enzymes and are grouped into the subclass of the peptide hydrolases or peptidases. Depending on the site of enzyme action the proteases can also be subdivided into exopeptidases or endopeptidases. Exopeptidases, such as aminopeptidases and carboxypeptidases catalyze the hydrolysis of the peptide bonds near the N– or C-terminal ends of the substrate, respectively (Fig. 4.5). Endopeptidases (Fig. 4.5) cleave peptide bonds at internal locations within the peptide sequence. Proteases may also be nonspecific and cleave all peptide bonds equally or they may be highly sequence specific and only cleave peptides after certain residues or within specific localized sequences.

The action of proteolytic enzymes is essential in many physiological processes. For example, proteases function in the digestion of food proteins, protein turnover, cell division, the blood-clotting cascade, signal transduction, processing of polypeptide hormones, apoptosis and the life-cycle of several disease-causing organisms including the replication of retroviruses such as the human immunodeficiency virus (HIV). Due to their key role in the life-cycle of many hosts and pathogens they have great medical, pharmaceutical, and academic importance.

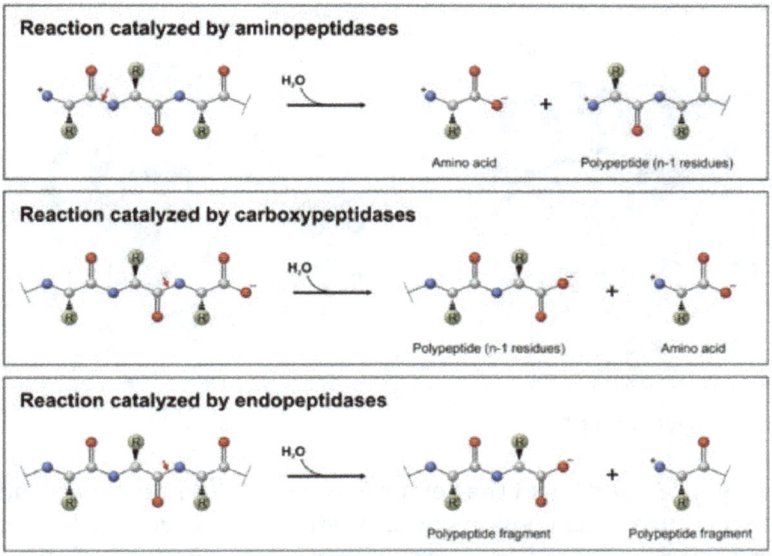

Fig. 4.5 Common Peptidase Reactions. Aminopeptidases (top diagram) and carboxypeptidases (middle diagram) remove the terminal amino acid residues, whereas endopeptidases (lower diagram) cleave protein sequences at internal sites. The red arrows show the peptide bonds to be cleaved.

It was estimated previously that about 2% of the human genes encode proteolytic enzymes and due to their necessity in many biological processes, proteases have become important therapeutic targets. They are intensively studied to explore their structure function relationships, to investigate their interactions with the substrates and inhibitors, to develop therapeutic agents for antiviral therapies or to improve their thermostability, efficiency and to change their specificity by protein engineering for industrial or therapeutic purposes.

Based on the catalytic mechanism and the presence of amino acid residue(s) at the active site the proteases can be grouped as aspartic proteases, cysteine proteases, glutamic proteases, metalloproteases, asparagine proteases, serine proteases, threonine proteases, and proteases with mixed or unknown catalytic mechanism. Here we will explore the reaction mechanism and sequence specificity of the family of serine proteases.

The list of serine proteases is quite long. They are grouped in two broad categories – 1) those that are chymotrypsin-like and 2) those that are subtilisin-like. Though subtilisin type and chymotrypsin-like enzymes use the same mechanism of action, including the catalytic triad, the enzymes are otherwise not related to each other by sequence and appear to have evolved independently (Figure 4.6). They are, thus, an example of convergent evolution – a process where evolution of different forms converge on a structure to provide a common function.

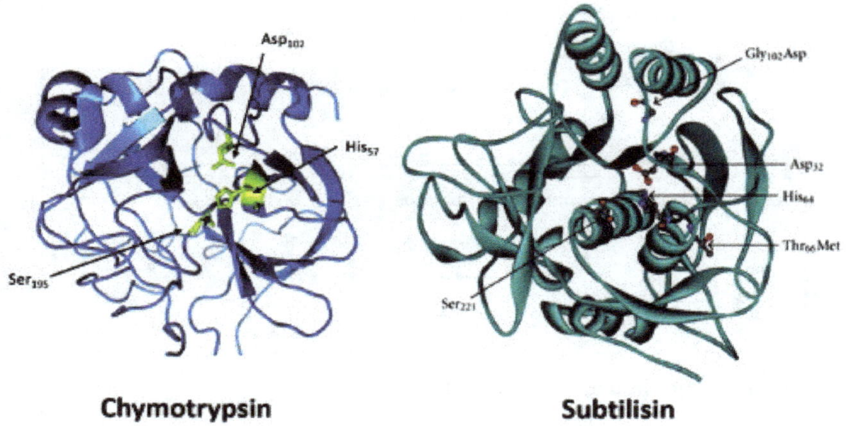

Fig.4.6 Convergent Evolution of the Serine Proteases, Chymotrypsin and Subtilisin. The crystal structure of the eukaryotic, Bovine Chymotrypsin (left hand panel) with catalytic triad indicated in green. The crystal structure of the prokaryotic Subtilisin BPN from the bacterium Bacillus subtilis (right hand panel) with the catalytic triad and common mutations indicated using the ball and stick models. Note that the catalytic triad formation is strikingly similar between the two structures, whereas the surrounding protein structures and sequence do not show homology or related ancestry.

The chymotrypsin-like serine protease enzymes cleave the peptide bond on the carboxylic acid side of specific amino acids and the specificity is determined by the size/ shape/charge of amino acid side chain that fits into the enzyme's S1 binding pocket (Figure 7.18). Three chymotrypsin-like family members that share high sequence homology are the pancreatic digestive enzymes, trypsin, chymotrypsin and elastase. The protein cleavage sites of these enzymes varies. Trypsin cleaves proteins on the carboxylic side of basic residues, such as lysine and arginine, while Chymotrypsin cleaves after aromatic hydrophobic amino acids, such as phenylalanine, tyrosine, and tryptophan, and Elastase cleaves after small, hydrophobic residues, such as glycine, alanine, and valine. As shown in Figure 4.7, variations in the amino acid residues within the binding pocket of these proteases, enables electrostatic interactions with the substrate and determines sequence specificity.

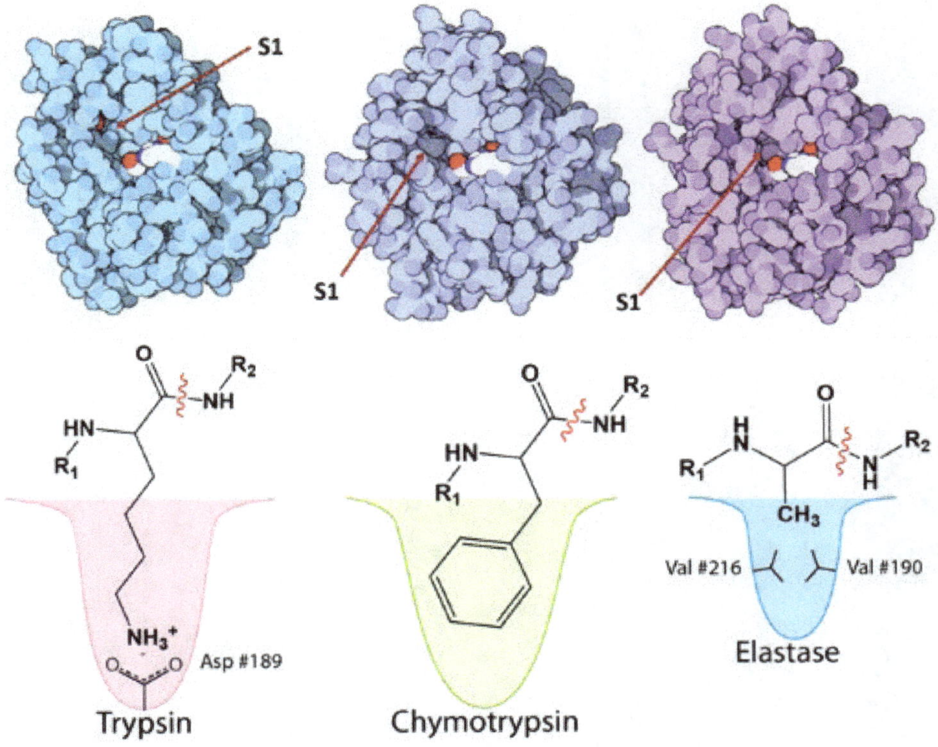

Fig. 4.7 Substrate Specificity of Trypsin, Chymotrypsin, and Elastase. The upper panel shows the space-filling crystal structures of Trypsin, Chymotrypsin, and Elastase, respectively, with the S1 substrate binding pocket indicated. The lower panel depicts the S1 binding domains of each protease in more detail with important amino acid R-groups indicated. For Trypsin, an aspartate residue in the lower portion of the S1 pocket aid in electrostatic interactions with basic residues of the substrate. The Chymotrypsin S1 binding pocket is large and hydrophobic in nature accommodating aromatic residues of the substrate, while the Elastase S1 binding pocket is small and hydrophobic, only allowing other small and hydrophobic R-groups to dock in this location.

5

CHEMICAL EQUILIBRIUM AND DYNAMIC EQUILIBRIA

CHEMICAL EQUILIBRIUM

Chemical equilibrium refers to the state of a system in which the concentration of the reactant and the concentration of the products do not change with time, and the system does not display any further change in properties.

When the rate of the forward reaction is equal to the rate of the reverse reaction, the state of chemical equilibrium is achieved by the system. When there is no further change in the concentrations of the reactants and the products due to the equal rates of the forward and reverse reactions, the system is said to be in a state of dynamic equilibrium.

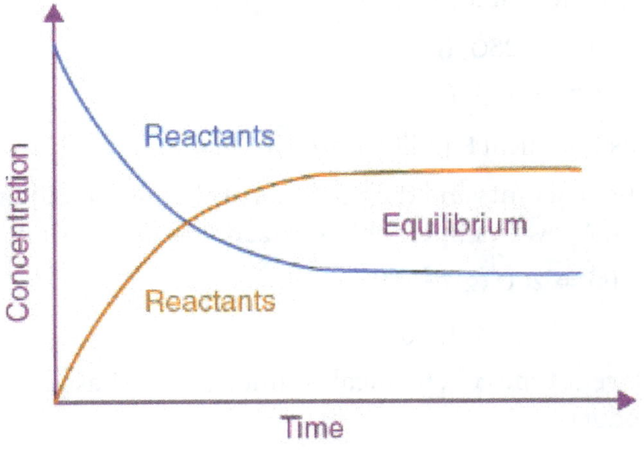

Fig. 5.1 State of Chemical Equilibrium

A graph (Fig. 5.1) with the concentration on the y-axis and time on the x-axis can be plotted. Once the concentration of both the reactants and the products stops showing change, chemical equilibrium is achieved.

Why Is Chemical Equilibrium Called Dynamic Equilibrium?

The stage at which the rate of the forward reaction is equal to the rate of the backward reaction is called an equilibrium stage. At this point, the number of reactant molecules converting into products and product molecules into reactants is the same. The same equilibrium can be carried out with the same reactants anywhere with similar conditions with continuous interchanging of molecules, and hence chemical equilibrium is dynamic.

EQUILIBRIUM

There are two types of chemical equilibrium:

- Homogeneous Equilibrium
- Heterogeneous Equilibrium

Homogenous Chemical Equilibrium

In this type, the reactants and the products of chemical equilibrium are all in the same phase. Homogenous equilibrium can be further divided into two types:

Reactions in which the number of molecules of the products is equal to the number of molecules of the reactants. For example,

- H_2 (g) + I_2 (g) ⇒ 2HI (g)
- N_2 (g) + O_2 (g) ⇒2NO (g)

Reactions in which the number of molecules of the products is not equal to the total number of reactant molecules. For example,

- $2SO_2$ (g) + O_2 (g) ⇒ $2SO_3$ (g)
- $COCl_2$ (g) ⇒ CO (g) + Cl_2 (g)

Heterogeneous Chemical Equilibrium

In this type, the reactants and the products of chemical equilibrium are present in different phases. A few examples of heterogeneous equilibrium are given below:

- CO_2 (g) + C (s) ⇒ 2CO (g)
- $CaCO_3$ (s) ⇒ CaO (s) + CO_2 (g)

Thus, the different types of chemical equilibrium are based on the phase of the reactants and products.

Factors Affecting Chemical Equilibrium

According to Le-Chatelier's principle, if there is any change in the factors affecting the equilibrium conditions, the system will counteract or reduce the effect of the overall transformation. This principle applies to both chemical and physical equilibrium.

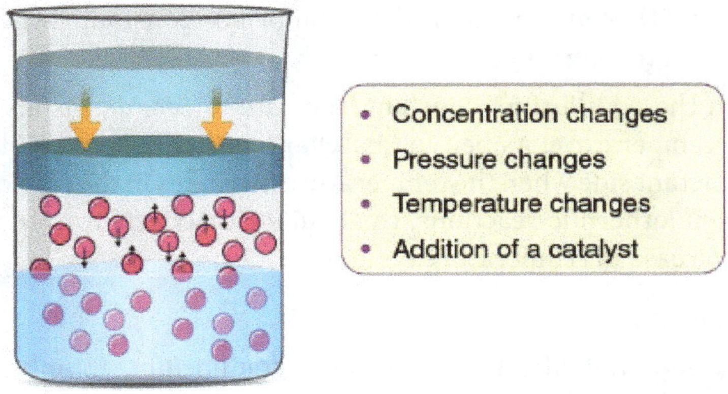

Fig. 5.2 Chemical Equilibrium – Factors Affecting Equilibrium

There are several factors like temperature, pressure and concentration of the system that affect equilibrium. Some important factors affecting chemical equilibrium are explained below.

Change in Concentration

- The concentration of the reactants or products added is relieved by the reaction, which consumes the substance that is added.
- The concentration of reactants or products removed is relieved by the reaction, which is in the direction that replenishes the substance which is removed.
- When the concentration of the reactant or product is changed, there is a change in the composition of the mixture in chemical equilibrium.

Effect of Change in Concentration

Change in Pressure

Change in pressure happens due to the change in the volume. If there is a change in pressure, it can affect the gaseous reaction as the total number of gaseous reactants and products are now different. According to Le Chatelier's principle, in heterogeneous chemical equilibrium, the change of pressure in both liquids and solids can be ignored because the volume is independent of pressure.

Effect of Change in Volume of Equilibrium Mixture

Change in Temperature

The effect of temperature on chemical equilibrium depends upon the sign of ÄH of the reaction and follows Le-Chatelier's Principle.

- As temperature increases, the equilibrium constant of an exothermic reaction decreases.
- In an endothermic reaction, the equilibrium constant increases with an increase in temperature.

Along with the equilibrium constant, the rate of reaction is also affected by the change in temperature. As per Le Chatelier's principle, the equilibrium shifts towards the reactant side when the temperature increases in the case of exothermic reactions; for endothermic reactions, the equilibrium shifts towards the product side with an increase in temperature.

Effect of a Catalyst

A catalyst does not affect the chemical equilibrium, it only speeds up a reaction. In fact, the catalyst equally speeds up the forward as well as the reverse reaction. This results in the reaction reaching its equilibrium faster. The same amount of reactants and products will be present at equilibrium in a catalysed or non-catalysed reaction. The presence of a catalyst only facilitates the reaction to proceed through a lower-energy transition state of reactants to products.

Effect of Addition of an Inert Gas

When an inert gas like argon is added to a constant volume, it does not take part in the reaction, so the equilibrium remains undisturbed. If the gas added is a reactant or product involved in the reaction, then the reaction quotient will change.

Effect of Addition of Inert Gas

In chemical reactions, reactants are converted into products by the forward reaction, and the products may be converted into reactants by the backward reaction. The two states, reactants and products, are different in composition. After some time of the start of the reaction, the rate of the forward and the backward reactions may become equal. Then, the number of reactants converted will be formed again by the reverse reaction such that the concentration of reactants and products do not change anymore. Hence, the reactants and products are in chemical equilibrium.

- $N2O_4 \Rightarrow 2NO_2$
- $PCl_5 \Rightarrow PCl_3 + PCl_2$
- $N_2 + H_2 \Rightarrow 2NH_3$

Importance of Chemical Equilibrium

It is useful in many industrial processes, like

- Preparation of ammonia by Haber's process: In this process, nitrogen combines with hydrogen to form ammonia; the yield of ammonia is more at low temperature, high pressure and in the presence of iron as a catalyst.
- Preparation of sulphuric acid by contact process: In this process, the fundamental reaction is the oxidation of sulphur dioxide into sulphur trioxide. This involves chemical equilibrium.

Dynamic Equilibrium

Dynamic Equilibrium can be defined as the state of a given system in which the reversible reaction taking place in it stops changing the ratio of reactants and products, but there is a movement of substances between the reactants and the products. This movement occurs at an equal rate, and there is no net change in the reactant and product ratio.

In other words, if we look at the fields of Physics and Chemistry, a dynamic equilibrium will mainly exist after the occurrence of a reversible reaction. The dynamic equilibrium in a reversible process will be found when the forward and reverse processes occur at the same rate. This will result in no observable change in the overall system. Once such a state is reached, the concentrations or partial pressures of all species will remain constant.

For these types of equilibria, the equilibrium constants are represented with the help of the rate constants for the forward and backward reactions. Systems maintaining a dynamic equilibrium are examples of systems in steady states.

Examples of Dynamic Equilibrium

A few important examples of dynamic equilibrium in our everyday life are listed below.

- A new bottle of an aerated drink has a specific value for the concentration of the carbon dioxide present in the liquid phase in it. When the bottle is opened, and half of the drink is poured out of it, the liquid carbon dioxide is slowly converted into gaseous carbon dioxide until a new point of equilibrium is reached, and the rate of conversion of CO2 from gas to liquid is equal to the rate of conversion of CO2 from liquid to the gaseous phase.
- The single-phase system in which acetic acid undergoes dissociation leads to an acid-base equilibrium. This state of dynamic equilibrium can be described by the following reaction:

$$CH_3COOH \Rightarrow CH_3COO- + H^+$$

- In the gaseous phase, it can be observed in the dimerization of nitrogen dioxide. Reaction: $2NO_2 \Rightarrow N2O_4$

- Henry's Law is applicable in the first example of dynamic equilibrium provided above, wherein the equilibrium concentration of carbon dioxide in the liquid phase is proportional to the partial pressure of the CO2 gas in the bottle.
- Industrial synthesis of ammonia via Haber's process.

Reaction: $N_2 (g) + 3H_2 (g) \Rightarrow 2NH_3 (g)$.

Difference between Static and Dynamic Equilibrium

Static equilibrium refers to a condition where the reaction occurring in a system is completely halted, and there exists no movement between the reactants and the products corresponding to the chemical reaction.

If the forces acting on an object cancel each other, in addition to the constancy of content and composition, no movement of the object takes place. This is static equilibrium.

The key differences between static and dynamic equilibrium are tabulated below.

Static vs Dynamic Equilibrium

Dynamic Equilibrium	Static Equilibrium
This type of equilibrium is reversible in nature.	This type of equilibrium is irreversible in nature.
This equilibrium implies that the reactants and the products are still participating in chemical reactions.	There is no further chemical reaction in the system.
In dynamic equilibrium, the forward and the backward reaction rates are equal	In static equilibrium, the forward and backward reaction rates are zero
It can only occur in closed systems	It can occur in both open and closed systems

However, the resultant force acting on both of these types of equilibria in a system is zero. Generally, neither of these types of equilibrium display visible changes.

EQUILIBRIUM CONSTANTS AND EXPRESSIONS

The equilibrium constant of a chemical reaction (usually denoted by the symbol K) provides insight into the relationship between the products and reactants when a chemical reaction reaches equilibrium. For example, the equilibrium constant of concentration (denoted by Kc) of a chemical reaction at equilibrium can be defined as the ratio of the concentration of products to the concentration of the reactants, each raised to their respective stoichiometric coefficients. It is important to note that there are different types of equilibrium constants that provide relationships between the products and the reactants of equilibrium reactions in terms of different units.

For a chemical reaction, the equilibrium constant can be defined as the ratio between the amount of reactant and the amount of product which is used to determine chemical behaviour.

At equilibrium, the rate of the forward reaction = rate of the backward reaction i.e., $r_f = r_b$ Or, $k_f \times a \times [A]^a[B]^b = k_b \times a \times [C]^c [D]^d$

At a particular temperature, the rate constants are constant. The ratio of the rate constant of forward reaction to the rate constant of backward reaction should be a constant and is called an equilibrium constant (Kequ).

Equilibrium Constant Formula

$$K_{equ} = k_f/k_b = [C]^c [D]^d/[A]a [B]^b = K_c$$

Where Kc indicates the equilibrium constant measured in moles per litre.

For reactions involving gases: The equilibrium constant formula, in terms of partial pressure, will be:

$$K_{equ} = k_f/k_b = [[pC]^c [pD]^d]/[[pA]^a [pB]^b] = K_p$$

Where Kp indicates the equilibrium constant formula in terms of partial pressures.

- Larger K_c/K_p values indicate higher product formation and higher percentage conversion.
- Lower K_c/K_p values indicate lower product formation and lower percentage conversion.
- Medium K_c/K_p values indicate optimum product formation.

Units of Equilibrium Constant

The equilibrium constant is the ratio of the concentrations raised to the stoichiometric coefficients. Therefore, the unit of the equilibrium constant = [Mole L-1]Dn.

Where, Δn = sum of stoichiometric coefficients of products – sum of stoichiometric coefficients of reactants.

Equilibrium Constant, Reaction Quotient and Gibbs Free Energy

K is the ratio of the relative amount of products to reactants at equilibrium, while Q is the ratio at any point in time of the reaction. The Q value can be compared to K to determine the direction of the reaction to take place. The spontaneity of the process is related to the free energy change. DG (Gibbs Free Energy), K (Equilibrium Constant), and Q (Reaction Quotient) are related, as detailed below:

- **$\Delta G < 0$ and $Q_c < K_c$ or K_p at the start of the reaction:** The reaction will proceed to form products.
- $\Delta G = 0$ and $Q_c < K_c$ or K_p at equilibrium, and no more changes in the concentration of the mixture.
- **$\Delta G > 0$ and $Q_c < K_c$ or K_p after equilibrium:** The reaction will proceed in the direction to form reactants.
- Kc = Equilibrium constant measured in moles per litre.
- Kp = Equilibrium constant calculated from the partial pressures

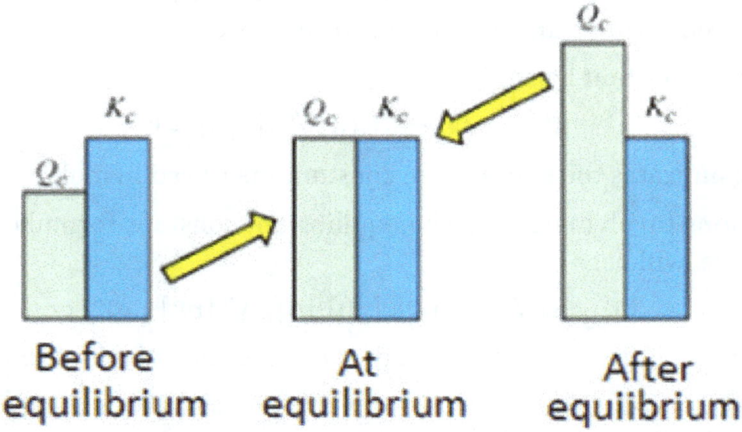

Fig. 5.3 Equilibrium Constant vs Reaction Quotient

Relationship between kc and k p

Consider the following reversible reaction: cC + dD Þ aA + bB

The equilibrium constant for the reaction is expressed in terms of the concentration (mole/litre):

$$\Rightarrow kc = [C]c[D]d[A]a[B]b$$

If the equilibrium involves gaseous species, then the concentrations are replaced by partial pressures of the gaseous substances. The equilibrium constant in terms of partial pressures is:

$$\Rightarrow kc = [pC]c[pD]d[pA]a[pB]b$$

Where pA, pB, pC and pD represent the partial pressures of the substance A, B, C and D, respectively. If gases are assumed to be ideal, then according to the ideal gas equation:

$$pV = nRT \text{ or } p = nRT/V$$

Where,

- p is the pressure in Pa
- n is the number of moles of gas
- V is the volume in m3
- T is the temperature in Kelvin
- n/V = molar concentration = [C]

Characteristics of Equilibrium Constant

- It is reaction specific, and at a constant temperature, it is fixed.
- A catalyst changes the rate of forward and backward reactions equally, not to affect the value of the equilibrium constant.
- Changes in concentration, pressure, temperature, and inert gases may affect the equilibrium, favouring either forward or backward reaction but not the equilibrium constant.
- It is related to the standard free energy as, ΔG0 = -RT ln K_{equ}.
- For the same reversible reaction, Kequ has different values at different temperatures.
- The equilibrium constant of the reverse equilibrium is the reciprocal of the original equilibrium, i.e. $K_{rev} = 1/K_{equ}$.
- If the equilibrium reaction stoichiometry is changed, the power of the equilibrium constant also gets changed by the same quantity.
- If the equilibrium constant for the reaction, A + B ⇒ C + D, is K, then, for equilibrium reaction 3A = 3B ⇒ 3C + 3D is K3.
- In case the stepwise multiple equilibria lead to the final products, the equilibrium constant of the net equilibrium = product of each stepwise equilibrium constant. Therefore, the net equilibrium constant $K = K_1 \times K_2 \times K_3$.
- Simultaneous equilibrium reactions have a common product. The equilibrium constant of the reactions does not change. Due to the higher concentration of the common product, the product concentrations will be reduced.

Applications of Equilibrium Constant

Equilibrium Constant for Predicting the Extent of Reaction

The equilibrium constant (K_c) can be used to predict the extent of a reaction, i.e. the degree of the disappearance of the reactants. The magnitude of the equilibrium constant gives an idea of the relative amount of the reactants and the products.

Case 1: The larger value of the equilibrium constant (>103) shows that forward reaction is favoured, i.e. the concentration of products is much larger than that of the reactants at equilibrium.

For example,

- $H_2(g) + Br_2(g) \Rightarrow 2HBr(g)$ Þ $Kc = 5.4\times10^{18}$
- $H_2(g) + Cl_2(g) \Rightarrow 2HCl(g)$ Þ $Kc = 4\times10^{31}$
- $H_2(g) + 12O_2(g) \Rightarrow H_2O(g)$ Þ $Kc = 2.4\times10^{47}$

This shows that at equilibrium, the concentration of the products is very high, i.e. reaction goes almost to completion.

Case 2: Intermediate value of the equilibrium constant (10-3 to 103) show that the concentration of the reactants and products are comparable.

For example,

- Fe_3 (aq) + SCN (aq) $\Rightarrow$ $[Fe(SCN)]_2$ (aq) $\Rightarrow$ Kc = 138 at 298 K
- H_2 (g) + I_2 (g) $\Rightarrow$ 2HI (g) $\Rightarrow$ Kc = 57 at 700 K.

Case 3: Low value of equilibrium constant (<10-3) shows that backward reaction is favoured, i.e. concentration of reactants is much larger than that of products, i.e. the reaction proceeds to a very small extent in the forward direction.

For example,

- N_2 (g) + O_2 (g) $\Rightarrow$ 2NO (g) $\Rightarrow$ $Kc = 4.8 \times 10^{-31}$ at 298K
- H_2O (g) $\Rightarrow$ H_2(g) + (1/2) O_2 (g) $\Rightarrow$ $Kc = 4.1 \times 10^{-48}$

Equilibrium Constant for Predicting the Direction of a Reaction

The equilibrium constant can be used to predict the direction of the reaction. We need a term, reaction quotient (Qc expressed in terms of concentrations or Qp in terms of partial pressures) similar to the equilibrium constant, except that the conditions are not at equilibrium.

For a balanced reaction, aA + bB ƒ cC + dD

The reaction quotient (Qc or Qp) is given as:

$$Q_c = [C]^c[D]^d/[A]^a[B]^b$$

$$Q_p = p^cC \times p^dD \,/\, p^aA \times p^bB$$

Comparison with Kc and the direction of reaction:

- If Q = K_c, the reaction is in equilibrium [Where, K_c = equilibrium constant] If Q > K_c, Q tends to decrease so as to become equal to K, the reaction will proceed in the backward direction.

- If $Q < K_c$, Q tends to increase so as to become equal to K, the reaction will proceed in the forward direction.

Factors Affecting Equilibrium Constant

Some factors that affect the equilibrium constant are as follows:

- Change in concentration of any product or reactant.
- Change in the pressure of the system.
- Change in temperature of the system.
- Adding inert gas.
- Adding catalyst.

THE EQUILIBRIUM CONSTANT IN RELATION TO REACTION QUOTIENTS

A scientist was observing a reaction and at a certain point and found the concentration of reactant is equal to the concentration of product and after some time and observed color of reactant is changing, the scientist found concentration of products is greater than the concentration of reactants, for example, it can be observed, the following reaction $N_2O_4 \Rightarrow 2NO_2$. Here initially the reaction was in colorless form and after some time when one observes the reaction was in reddish-brown in color and wants to know what is actually happening and to know this scientists invented a new concept called equilibrium constant and here are some terms we need to know about the equilibrium constant. The rate of reaction is the speed at which a chemical reaction takes place.

Equilibrium

It is the condition of a reversible chemical reaction in which there is no net change in the amount of reactant and product. At equilibrium rate of forwarding reaction = rate of backward reaction. Equilibrium outside equilibrium is recognized by the consistency of certain measurable properties like Intensity of color, density, concentration, and Pressure. Equilibrium is dynamic in nature but not static.

Equilibrium Constant

It is the ratio between the product of the concentration of products raised to their powers and the product of the concentration of reactants raised to their powers at equilibrium.

The Equilibrium Constant (K)

In 1864, the Norwegian chemists Cato Guldberg (1836–1902) and Peter Waage (1833– 1900) carefully measured the compositions of many reaction systems at equilibrium. They discovered that for any reversible reaction of the general form

$$m A + n B \Rightarrow x C + y D$$

where A and B are reactants, C and D are products, and m, n, x, and y are the stoichiometric coefficients in the balanced chemical equation for the reaction, the ratio of the product of the equilibrium quantities of the products (raised to their coefficients in the balanced chemical equation) to the product of the equilibrium concentrations of the reactants (raised to their coefficients in the balanced chemical equation) is always a constant under a given set of conditions. This relationship was eventually summarized as follows:

$$K = \frac{(a_C)^x (a_D)^y}{(a_A)^m (a_B)^n}$$

Eq 1 Equilibrium Constant

where K is the equilibrium constant for the reaction, equivalent to the value defined in Section 4.1, and aX represents the activity of each species participating in the equilibrium. The chemical equilibrium equation represented with reactants A & B and products C & D is called the equilibrium equation, and the right side of the mathematical equation above is called the equilibrium constant expression. The relationship shown in the expression for K is true for any pair of opposing reactions regardless of the mechanism of the reaction or the number of steps in the mechanism.

An important fact to note is that equilibrium constants are dimensionless (they have no units) but the temperature at which this value is valid must always be listed (since K is temperature dependent). This is due to computing K values using the activities of the reactants and products in the equilibrium system. The activity of a substance is a measure of its effective concentration under specified conditions. While a detailed discussion of this important quantity is beyond the scope of an introductory text, it is necessary to be aware of a few important aspects:

- Activities are dimensionless (unitless) quantities and are in essence "adjusted" quantities of reactants and products.
- For relatively dilute solutions, a solute's activity and its molar concentration are roughly equal (i.e. for a solute X, a X H» [X] in mol/L or M). Note that this approximation does not hold for highly concentrated solutes.
- For gases, a substance's activity is equal to its partial pressure (i.e. a X = P X in bar)
- Activities for pure condensed phases (solids and liquids) are equal to 1 (i.e. a X = 1), hence their activities do not appear in the expression for K

Further discussion about activities can be found here.

We categorize equilibria into two types: homogeneous and heterogeneous. A homogeneous equilibrium is one in which all of the reactants and products are present in a single solution (by definition, a homogeneous mixture). In this chapter, we will concentrate on the two most common types of homogeneous equilibria: those occurring in liquid-phase solutions and those involving exclusively gaseous species. A heterogeneous equilibrium is a system in which reactants and products are found in two or more phases. The phases may be any combination of solid, liquid, or gas phases, and solutions. When dealing with these equilibria, remember that solids and pure liquids do not appear in equilibrium constant expressions (as we've mentioned above, the activities of pure solids, pure liquids, and solvents are 1).

Derivation of Equilibrium Constant

According to the law of mass action, the rate of any chemical reaction at a given temperature is directly proportional to the product of the molar concentration of reactants rising to a number of moles in their products. [Law of mass action is applicable to reactions that are at chemical equilibrium only For physical changes the law of mass action cannot be applied]. Suppose if the reaction is, aA + bB â! cC + dD. Now let's apply the law of mass action to forward reaction rf is proportional to [A]a[B]b (Here f indicates the rate for forward reaction).

$$r_f = k_f[A]^a[B]^b$$

(Here kf is the rate constant for forwarding reaction). Now let's apply the law of mass action to backward reaction, rb is proportional to [C]c[D]d

(Here b indicates the rate for backward reaction)

$$r_b = k_b[C]^c[D]^d$$

(Here kb is rate constant for backward reaction)

At equilibrium: Rate of forward reaction = rate for backward reaction kf[A]a[B] b = kb[C]c[D]d

$$\text{So, } k_f/k_b = [C]^c[D]^d/[A]^a[B]^b$$

[Note: kf/kb = kc] and kc is equilibrium constant in terms of concentration kc = (Product of concentration of product raised to powers)/(product of the concentration of reactant raised to powers).

The Reaction Quotient, Q

We previously saw that knowing the magnitude of the equilibrium constant under a given set of conditions allows chemists to predict the extent of a reaction. Often, however, chemists must decide whether a system has reached equilibrium or if the composition of the mixture will continue to change with time.

To determine whether a system has reached equilibrium, chemists use a quantity called the reaction quotient (Q). The expression for the reaction quotient has precisely the same form as the equilibrium constant expression, except that Q may be derived from a set of values measured at any time during the reaction of any mixture of the reactants and products, regardless of whether the system is at equilibrium. Therefore, for the following general reaction:

m A + n B Ì! x C + y D

the reaction quotient is defined as follows:

$$Q = \frac{(a_C)^x (a_D)^y}{(a_A)^m (a_B)^n}$$

Eq 2 Reaction Quotient

Similarly to the equilibrium constant, the reaction quotient is dimensionless (no units) - this stems from using the species activities as its effective concentrations. As before, the activity of each species participating in the equilibrium can be represented as follows:

Chemical Equilibrium and Dynamic Equilibria 73

- For a solute X, a_X H" [X] in mol/L (note that again, this does not apply to highly concentrated solutions)
- gases, $a_X = P_X$ in bar
- For pure solids and liquids, $a_X = 1$

SHIFTING EQUILIBRIA AND EQUILIBRIUM CALCULATIONS

The convention for writing chemical equations involves placing reactant formulas on the left side of a reaction arrow and product formulas on the right side. By this convention, and the definitions of "reactant" and "product," a chemical equation represents the reaction in question as proceeding from left to right. Reversible reactions, however, may proceed in both forward (left to right) and reverse (right to left) directions. When the rates of the forward and reverse reactions are equal, the concentrations of the reactant and product species remain constant over time and the system is at equilibrium. The relative concentrations of reactants and products in equilibrium systems vary greatly; some systems contain mostly products at equilibrium, some contain mostly reactants, and some contain appreciable amounts of both.

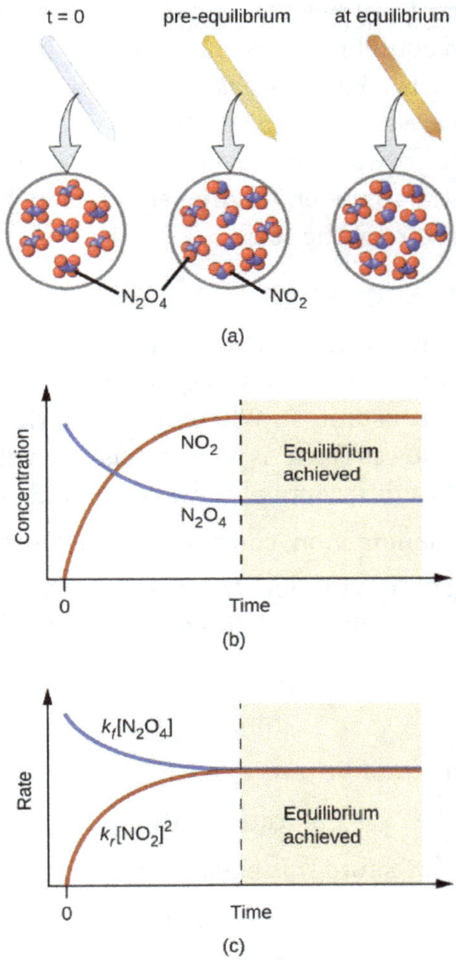

Figure 5.4 (a) A sealed tube containing colorless N2O4 darkens as it decomposes to yield brown NO2. (b) Changes in concentration over time as the decomposition reaction achieves equilibrium. (c) At equilibrium, the forward and reverse reaction rates are equal.

For this elementary process, rate laws for the forward and reverse reactions may be derived directly from the reaction stoichiometry:

$$ratef=kf[N2O_4]$$

$$rater=kr[NO_2]$$

As the reaction begins (t = 0), the concentration of the N2O4 reactant is finite and that of the NO2 product is zero, so the forward reaction proceeds at a finite rate while the reverse reaction rate is zero. As time passes, N2O4 is consumed and its concentration falls, while NO2 is produced and its concentration increases (b). The decreasing concentration of the reactant slows the forward reaction rate, and the increasing product concentration speeds the reverse reaction rate (c). This process continues until the forward and reverse reaction rates become equal, at which time the reaction has reached equilibrium, as characterized by constant concentrations of its reactants and products (shaded areas of [b and c). It's important to emphasize that chemical equilibria are dynamic; a reaction at equilibrium has not "stopped," but is proceeding in the forward and reverse directions at the same rate. This dynamic nature is essential to understanding equilibrium behavior as discussed in this and subsequent chapters of the text.

Effect of a Change in Temperature

Consistent with the law of mass action, an equilibrium stressed by a change in concentration will shift to re-establish equilibrium without any change in the value of the equilibrium constant, K. When an equilibrium shifts in response to a temperature change, however, it is re-established with a different relative composition that exhibits a different value for the equilibrium constant.

To understand this phenomenon, consider the elementary reaction A Ì! B

Since this is an elementary reaction, the rates laws for the forward and reverse may be derived directly from the balanced equation's stoichiometry:

$$rate_f = k_f[A]$$

$$rate_r = k_r[B]$$

When the system is at equilibrium,

$$rate_r = rate_f$$

Substituting the rate laws into this equality and rearranging gives

$$k_f[A]\ k_r[B]$$

$$[B]/[A]=Kf/Kr = K_c$$

The equilibrium constant is seen to be a mathematical function of the rate constants for the forward and reverse reactions. Since the rate constants vary with temperature as described by the Arrhenius equation, is stands to reason that the equilibrium constant will likewise vary with temperature (assuming the rate constants are affected to different extents by the temperature change). For more complex reactions involving multistep reaction mechanisms, a similar but more complex mathematical relation exists between the equilibrium constant and the rate constants of the steps in the mechanism. Regardless of how complex the reaction may be, the temperature-dependence of its equilibrium constant persists.

Predicting the shift an equilibrium will experience in response to a change in temperature is most conveniently accomplished by considering the enthalpy change of the reaction. For example, the decomposition of dinitrogen tetroxide is an endothermic (heat-consuming) process:

$$N_2O_4(g) \Rightarrow 2NO_2(g)\ \Delta H = +57.20\ kJ$$

For purposes of applying Le Chatelier's principle, heat (q) may be viewed as a reactant:

$$heat + N_2O_4(g) \Rightarrow 2NO_2(g)$$

Raising the temperature of the system is akin to increasing the amount of a reactant, and so the equilibrium will shift to the right. Lowering the system temperature will likewise cause the equilibrium to shift left. For exothermic processes, heat is viewed as a product of the reaction and so the opposite temperature dependence is observed.

Effect of a Catalyst

The kinetics chapter of this text identifies a catalyst as a substance that enables a reaction to proceed via a different mechanism with an accelerated rate. The catalyzed reaction mechanism involves a lower energy transition state than the uncatalyzed reaction, resulting in a lower activation energy, Ea, and a correspondingly greater rate constant.

To discern the effect of catalysis on an equilibrium system, consider the reaction diagram for a simple one-step (elementary) reaction shown in Figure 5.5. The lowered transition state energy of the catalyzed reaction results in lowered activation energies for both the forward and the reverse reactions. Consequently, both forward and reverse reactions are accelerated, and equilibrium is achieved more quickly but without a change in the equilibrium constant.

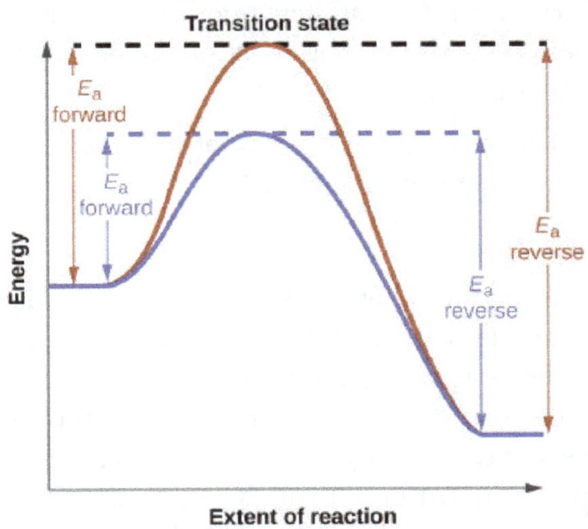

Fig. 5.5 Reaction diagrams for an elementary process in the absence (red) and presence (blue) of a catalyst. The presence of catalyst lowers the activation energies of both the forward and reverse reactions but does not affect the value of the equilibrium constant.

An interesting case study highlighting these equilibrium concepts is the industrial production of ammonia, NH3. This substance is among the "top 10" industrial chemicals with regard to production, with roughly two billion pounds produced annually in the US. Ammonia is used as a chemical feedstock to synthesize a wide range of commercially useful compounds, including fertilizers, plastics, dyes, and explosives.

Most industrial production of ammonia uses the Haber-Bosch process based on the following equilibrium reaction:

N2(g) + 3H2(g) ƒ 2NH3(g) DH = "92.2 kJ

The traits of this reaction present challenges to its use in an efficient industrial process. The equilibrium constant is relatively small (Kp on the order of 10"5 at 25 °C), meaning very little ammonia is present in an equilibrium mixture. Also, the rate of this reaction is relatively slow at low temperatures. To raise the yield of ammonia, the industrial process is designed to operate under conditions favoring product formation:

- High pressures (concentrations) of reactants are used, ~150"250 atm, to shift the equilibrium right, favoring product formation.
- Ammonia is continually removed (collected) from the equilibrium mixture during the process, lowering its concentration and also shifting the equilibrium right.
- Although low temperatures favor product formation for this exothermic process, the reaction rate at low temperatures is inefficiently slow. A catalyst is used to accelerate the reaction to reasonable rates at relatively moderate temperatures (400"500 °C).

A diagram illustrating a typical industrial setup for production of ammonia via the Haber-Bosch process is shown in Figure 5.6.

Equilibrium Calculations

Having covered the essential concepts of chemical equilibria in the preceding sections of this chapter, this final section will demonstrate the more practical aspect of using these concepts and appropriate mathematical strategies to perform various equilibrium calculations. These types of computations are essential to many areas of science and technology—for example, in the formulation and dosing of pharmaceutical products. After a drug is ingested or injected, it is typically involved in several chemical equilibria that affect its ultimate concentration in the body system of interest. Knowledge of the quantitative aspects of these equilibria is required to compute a dosage amount that will solicit the desired therapeutic effect.

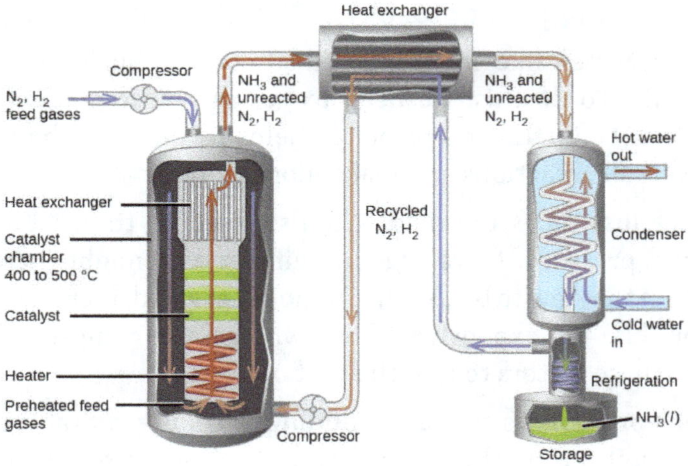

Figure 5.6 The figure shows a typical industrial setup for the commercial production of ammonia by the Haber- Bosch process. The process operates under conditions that stress the chemical equilibrium to favor product formation.

Many of the useful equilibrium calculations that will be demonstrated here require terms representing changes in reactant and product concentrations. These terms are derived from the stoichiometry of the reaction, as illustrated by decomposition of ammonia:

2NH3(g)⇒ N2(g)+3H2(g)

As shown earlier in this chapter, this equilibrium may be established within a sealed container that initially contains either NH3 only, or a mixture of any two of the three chemical species involved in the equilibrium. Regardless of its initial composition, a reaction mixture will show the same relationships between changes in the concentrations of the three species involved, as dictated by the reaction stoichiometry. For example, if the nitrogen concentration increases by an amount x:

D[N2]=+x

APPLICATION OF LE CHATELIER'S PRINCIPLE

Le Chatelier's Principle is a crucial tool for businesses in industry. As a business, you want to make as much product as possible while keeping costs low to make the most money. By using Le Chatelier's Principle, we can change the conditions of a reaction to make it produce more of the product we want. This is called finding the compromise conditions. In this article, we'll talk about how Le Chatelier's Principle is used in real life situations. We'll look at four examples of how it's used to maximise yield and profits.

The importance of applications of Le Chatelier's principle

Industrial processes like making fertiliser or sulphuric acid are all about maximizing profit. To do this, we need to get the most product we can while keeping costs low. Le Chatelier's principle helps us increase the yield of certain products in reversible reactions and make more money.

But we can't just focus on yield. We also need to think about factors like temperature and pressure. Lowering the temperature might increase yield but make the reaction too slow to be useful. On the other hand, increasing temperature or pressure could be too expensive. That's why we use compromise conditions - they balance all these factors to find the best option.

Le Chatelier's principle is crucial for finding the most profitable combination of reactants and conditions. Without it, our industrial processes would be much less efficient. In this article, we'll look at some real-life examples of how Le Chatelier's principle is used in industry.

Applications of Le Chatelier's Principle in the chemical industry

Now let's dive into some real-life examples of how Le Chatelier's principle is used in industry. We'll explore four different compounds:

- Methanol
- Ethanol
- Sulphuric acid (made using the Contact process)
- Ammonia (made using the Haber process)

Here's a breakdown of how each of these compounds is made using Le Chatelier's principle.

Le Chatelier's principle and methanol production

Methanol is made by reacting synthesis gas, which is a mixture of carbon monoxide and hydrogen, with a copper catalyst. It has the following equation:

$$CO(g) + 2H2(g) \Rightarrow CH3OH(g)\ \Delta H = -91\ kJ\ mol\text{-}1$$

You should now be able to predict the effect of certain conditions on the yield of methanol:

The forward reaction is exothermic. This means a lower temperature shifts the equilibrium to the right and increases the yield of methanol. However, a temperature that is too low slows down the rate of reaction and so a compromise temperature of 500 K is used.The forward reaction produces fewer moles of gas. This means that increasing the pressure shifts the equilibrium to the right and increases the yield of methanol. However, maintaining a high pressure is expensive, and so a compromise pressure of 10,000 kPa is used.

33 million tonnes of methanol are produced every year. Most of it is used to make methanal, an aldehyde further transformed into many types of plastics. However, methanol is also seeing a surge in popularity as a fuel. It can be used in typical diesel and petrol cars with little modification to their existing engines and is even being tested in boats.

Next, we'll take a look at making another alcohol, ethanol.

Le Chatelier's principle and ethanol production

When it comes to making ethanol using the hydration of ethene, Le Chatelier's principle plays a crucial role in maximizing yield while keeping costs in check.

The reaction is exothermic, meaning that a lower temperature will increase the yield of ethanol. However, too low a temperature will slow down the reaction too much, so a compromise temperature of 570 K is used.

The forward reaction produces fewer moles of gas, so increasing pressure will shift the equilibrium to the right and increase the yield of ethanol. However, maintaining high pressure is expensive, so a compromise pressure of 6,500 kPa is used.

Adding excess steam also increases the yield of ethanol, but too much steam dilutes the catalyst and slows down the reaction. Instead, ethanol is removed as it is formed, decreasing its concentration and favoring the forward reaction. Ethene and steam are then repeatedly recycled over the catalyst.

Overall, Le Chatelier's principle helps find the balance between yield and cost in the production of ethanol using the hydration of ethene. Ethanol is not only used in alcoholic drinks but also plays a crucial role as an antimicrobial agent, destroying microorganisms by disrupting their lipid bilayer membrane and denaturing their proteins

Le Chatelier's principle and the Contact process

Le Chatelier's principle plays a crucial role in the industrial production of various compounds, such as methanol, ethanol, sulphuric acid, and ammonia.

In the production of ethanol using the hydration of ethene, Le Chatelier's principle helps find the balance between yield and cost by adjusting temperature, pressure, and steam concentration. The same principle is applied in the Contact process for the reversible reaction of sulphur dioxide to sulphur trioxide. By adjusting temperature, pressure, and oxygen concentration, the yield of sulphur trioxide can be increased.

The Haber process for ammonia production also uses Le Chatelier's principle to increase yield by adjusting temperature and pressure. In all of these processes, product is removed, and unreacted gases are recycled to increase yield.

Overall, Le Chatelier's principle helps industries find a balance between yield and cost, allowing for the efficient production of various compounds used in fertilizers, detergents, pharmaceuticals, and even explosives.

Illustrations of le chatelier's principle

Color changes of Cobalt(II) chloride

The CoCl2.6H2O or [Co(H2O)6]Cl2 is a deep purple colored solid. It forms a purple colored solution when is dissolved in water. The color of [Co(H2O)6]2+ ion is pink. However it exists in equilibrium with small amount of [CoCl4]2- that is intense blue in color. The equilibrium is shown below. The purple color of the solution is the result of combination of these two colors.

The purple color of the aqueous solution of Cobalt(II) chloride can be changed by adjusting different variables in the following experiments.

$$\underset{\text{pink}}{[Co(H_2O)_6]^{2+}_{(aq)}} + 4Cl^-_{(aq)} + 50kJ \rightleftharpoons \underset{\text{intense blue}}{[CoCl_4]^{2-}_{(aq)}} + 6H_2O_{(aq)}$$

- The color of the solution turns intense blue upon addition of conc. HCl to the first test tube.

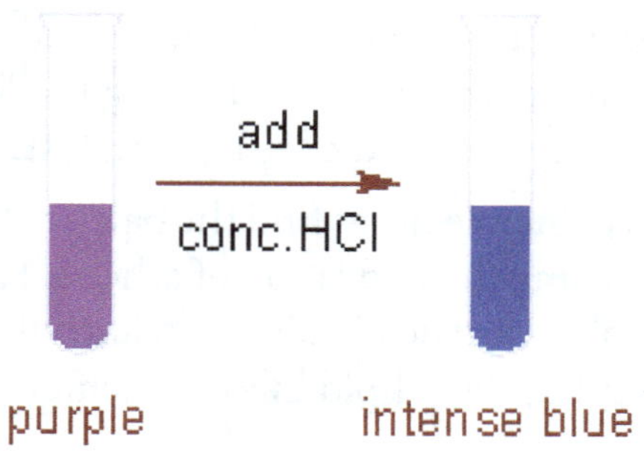

This is because of increase in the concentration of Cl- ions, which are furnished by HCl. As a result, the forward reaction is favored to give more [CoCl4]2-. Hence the color is turned to intense blue.

The Cl- ions are common to both HCl and [Co(H2O)6]Cl2. Hence this is also referred to as "common ion effect".

- The color of the solution turns to pale pink by adding excess of water in the second experiment.

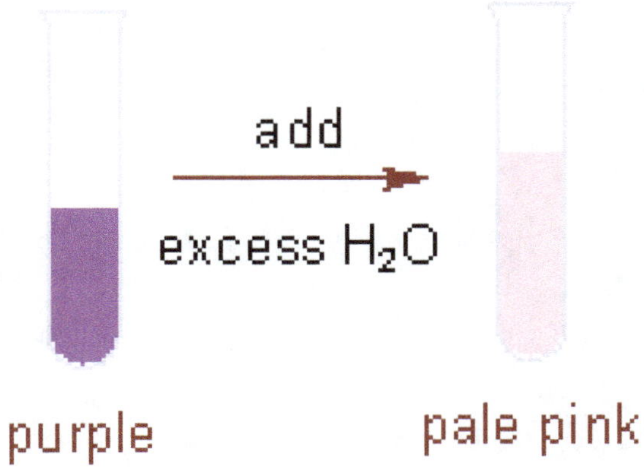

According to le Chatelier's principle, the backward reaction is favored when the concentration of one of the product increases. Hence increase in the concentration of water favors backward reaction. That will result in the formation of more [Co(H2O)6]2+, which is pale pink in color.

- In the third experiment, the color turns to blue upon heating the purple colored solution.

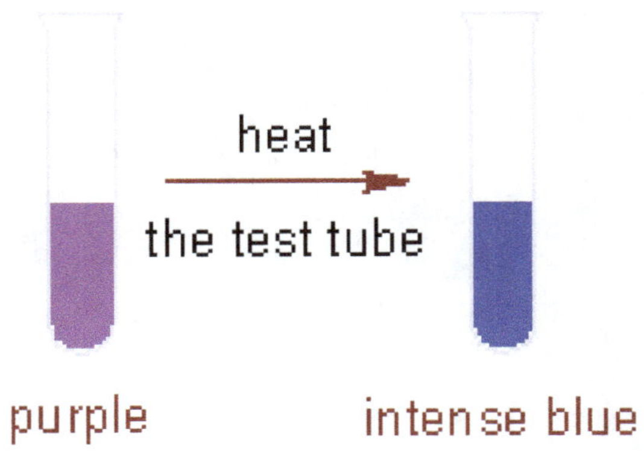

According to le Chatelier's principle, when heat is added to the system, the endothermic reaction is favored to remove heat from the system. In this case, the formation of [CoCl4]2- is an endothermic reaction. Hence more amount of blue colored [CoCl4]2- is formed.

6

ACIDS AND BASES IN CHEMICAL REACTIVITY

The definition of an acid is often cited as: any compound that increases the amount of hydrogen ion (H+) in an aqueous solution. The chemical opposite of an acid is a base. The equivalent definition of a base is that a base is a compound that increases the amount of hydroxide ion (OH") in an aqueous solution. These original definitions were proposed by Arrhenius (the same person who proposed ion dissociation) in 1884, so they are referred to as the Arrhenius definition of an acid and a base, respectively.

You may recognize that, based on the description of a hydrogen atom, an H+ ion is a hydrogen atom that has lost its lone electron; that is, H+ is simply a proton. Do we really have bare protons moving about in aqueous solution? No. What is more likely is that the H+ ion has attached itself to one (or more) water molecule(s). To represent this chemically, we define the hydronium ion H3O+(aq), a water molecule with an extra hydrogen ion attached to it. as H3O+, which represents an additional proton attached to a water molecule. We use the hydronium ion as the more logical way a hydrogen ion appears in an aqueous solution, although in many chemical reactions H+ and H3O+ are treated equivalently.

For purposes of this brief introduction, we will consider only the more common types of acid-base reactions that take place in aqueous solutions. In this context, an acid is a substance that will dissolve in water to yield hydronium ions, H3O+. As an example, consider the equation shown here:

$$HCl(aq)+H_2O(aq) \rightarrow Cl\text{-}aq)+H_3O+(aq)$$

The process represented by this equation confirms that hydrogen chloride is an acid. When dissolved in water, H3O+ ions are produced by a chemical reaction in which H+ ions are transferred from HCl molecules to H2O molecules (Figure 6.1).

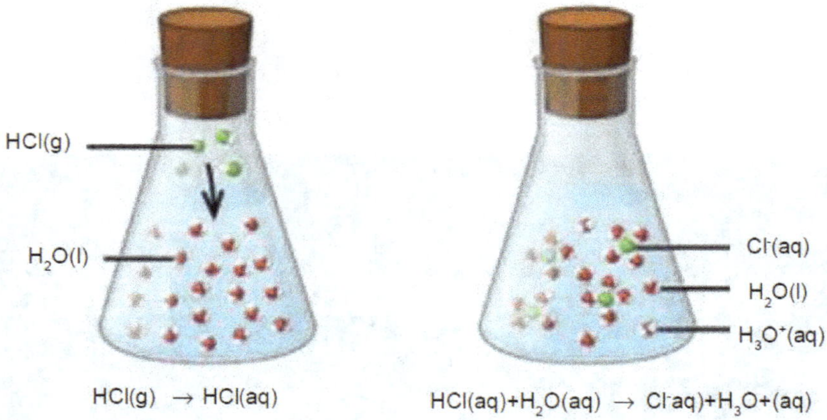

Fig. 6.1. When hydrogen chloride gas dissolves in water, (a) it reacts as an acid, transferring protons to water molecules to yield (b) hydronium ions (and solvated chloride ions).

The nature of HCl is such that its reaction with water as just described is essentially 100% efficient: Virtually every HCl molecule that dissolves in water will undergo this reaction. Acids that completely react in this fashion are called strong acids, and HCl is one among just a handful of common acid compounds that are classified as strong (Table 6.1).

Table 6.1 Common Strong Acids

Compound Formula	Name in Aqueous Solution
HBr	hydrobromic acid
HCl	hydrochloric acid
HI	hydroiodic acid
HNO_3	nitric acid
$HClO_4$	perchloric acid
$HClO_3$	chloric acid
H_2SO_4	sulfuric acid

A far greater number of compounds behave as weak acids and only partially react with water, leaving a large majority of dissolved molecules in their original form and generating a relatively small amount of hydronium ions.

Table 6.2. Common Weak Acids

Compound Formula	Name in Aqueous Solution
HF	hydrofluoric acid
HCN	hydrocyanic acid
$HC_2H_3O_2$	acetic acid
HNO_2	nitrous acid

Compound Formula	Name in Aqueous Solution
HClO	hypochlorous acid
$HClO_2$	chlorous acid
H_2SO_3	sulfurous acid
H_2CO_3	carbonic acid
H_3PO_4	phosphoric acid

Weak acids are commonly encountered in nature, being the substances partly responsible for the tangy taste of citrus fruits, the stinging sensation of insect bites, and the unpleasant smells associated with body odor. A familiar example of a weak acid is acetic acid, the main ingredient in food vinegars:

$$CH_3CO_2H(aq) + H_2O(l) \Rightarrow CH3CO2-(aq)+H_3O+(aq)$$

When dissolved in water under typical conditions, only about 1% of acetic acid molecules are present in the ionized form, CH3CO2– (Fig. 6.2). The use of a double-arrow in the equation above denotes the partial reaction aspect of this process, a concept addressed fully in the chapters on chemical equilibrium.)

A base is a substance that will dissolve in water to yield hydroxide ions, OH". The most common bases are ionic compounds composed of alkali or alkaline earth metal cations (groups 1 and 2) combined with the hydroxide ion—for example, NaOH and Ca(OH)2. When these compounds dissolve in water, hydroxide ions are released directly into the solution. For example, KOH and Ba(OH)2 dissolve in water and dissociate completely to produce cations (K+ and Ba2+, respectively) and hydroxide ions, OH". These bases, along with other hydroxides that completely dissociate in water, are considered strong bases.

Consider as an example the dissolution of lye (sodium hydroxide) in water:

$$NaOH(s) \longrightarrow Na+(aq)+OH-(aq)$$

This equation confirms that sodium hydroxide is a base. When dissolved in water, NaOH dissociates to yield Na+ and OH" ions. This is also true for any other ionic compound containing hydroxide ions. Since the dissociation process is essentially complete when ionic compounds dissolve in water under typical conditions, NaOH and other ionic hydroxides are all classified as strong bases.

(a) (b)

Fig. 6.2. (a) Fruits such as oranges, lemons, and grapefruit contain the weak acid citric acid. (b) Vinegars contain the weak acid acetic acid.

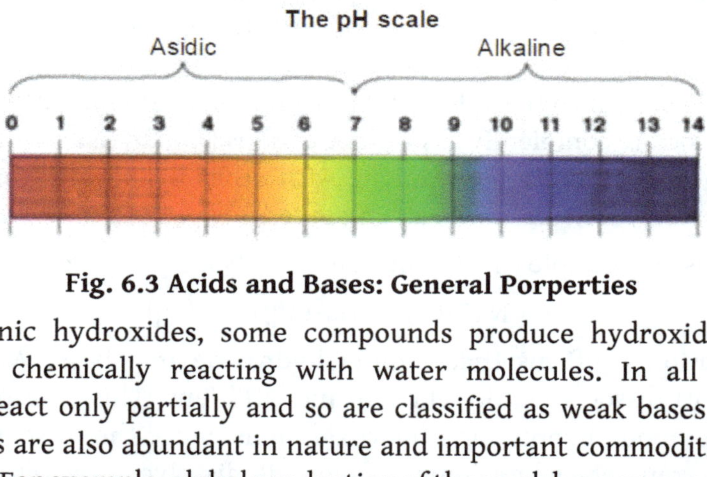

Fig. 6.3 Acids and Bases: General Porperties

Unlike ionic hydroxides, some compounds produce hydroxide ions when dissolved by chemically reacting with water molecules. In all cases, these compounds react only partially and so are classified as weak bases. These types of compounds are also abundant in nature and important commodities in various technologies. For example, global production of the weak base ammonia is typically well over 100 metric tons annually, being widely used as an agricultural fertilizer, a raw material for chemical synthesis of other compounds, and an active ingredient in household cleaners (Fig. 6.3). When dissolved in water, ammonia reacts partially to yield hydroxide ions, as shown here:

$$NH_3(aq)+H_2O(l) \Rightarrow NH4+(aq)+OH-(aq)$$

Under typical conditions, only about 1% of the dissolved ammonia is present as NH4+ ions.

Physical Properties of Acid

- The word "acid" comes from the Latin word 'acere' which means sour. This distinguishable property helps identify acids from other compounds such as salt and bases. Many acids can be hazardous if ingested and shouldn't be tasted.
- Once the acid binds to the base, it becomes a neutral substance. Often this reaction can lead to water and salt. This is often seen when strong acids react with strong bases.
- Acids in an aqueous solution produce hydrogen ions which are responsible for the conductivity of the solution. The acid that conducts electricity strongly is a strong acid, and the acid that conducts electricity weakly is a weak acid.

Chemical Properties of Acid and Bases

Reactions of Acids and Bases with Metals

When a metal reacts with an acid, it generally displaces hydrogen from the acids. This leads to the evolution of hydrogen gas. The metals combine with the remaining part of acids to form a salt. For example, the reaction of sulphuric acid with zinc.

$$H_2SO_4 + Zn \rightarrow ZnSO_4 + H_2$$

Alkalis (bases that are soluble in water) react with metals to produce salt and hydrogen gas. For example, reaction of zinc with sodium hydroxide.

Acids and Bases in Chemical Reactivity

$$2NaOH + Zn \rightarrow Na_2ZnO_2 + H_2$$

The Reaction of Metal Carbonates/Metal Bicarbonates with Acids

Metal carbonates/metal bicarbonates react with acids to produce salt, carbon dioxide and water. For example the reaction of sodium carbonate/sodium bicarbonate with hydrochloric acid.

$$Na_2CO_3 + HCl\,(aq) \rightarrow 2NaCl(aq) + H_2O(l) + CO_2$$

The Reaction of Metal Oxide with Acids

Metal oxides react with acids to produce salt and water. For example reaction of copper oxide and dilute hydrochloric acid.

$$CuO\ 2HCl \rightarrow CuCl_2 + H_2$$

The Reaction of Non-metal Oxide with Bases

Non-metal oxides react with bases to produce salt and water. For example the reaction of carbon dioxide and lime water (calcium hydroxide)

$$CO_2 + Ca(OH)_2 \rightarrow CaCO_3 + H_2O$$

The Reaction between Acids and Bases

Acids react with bases to produce salt and water. The reaction between acids and bases to give salts is known as neutralization reactions. For example the reaction of sodium hydroxide with hydrochloric acid.

$$NaOH + HCl \rightarrow NaCl + H_2O$$

Neutralization of Acid and Base

The reaction between an acid and a base invariably gives salt and water and is called neutralization. In a neutralization reaction, one H+ ion of acid is neutralized by one OH– ion of the base. When all the H+ ions in the acidic solution are neutralized by the same number of OH– ions of the basic solution, it is called complete neutralization. The relative amounts of acid and base required for complete neutralization depend upon the total number of H+ and OH– ions produced by the respective acid and base.

BRONSTED-LOWRYANDLEWISDEFINITIONSOFACIDSANDBASES

BRONSTED ACID

According to Bronsted, an acid is a substance capable of donating proton or H+ to another compound and forming a conjugate base.

- Bronsted regarded acid as a proton donor.

Consider a reaction,

$$HCl(aq)+NH3(aq) \rightarrow NH_4+(aq)+Cl''(aq)$$

Here, hydrochloric acid (HCl) donates its proton (H^+) to ammonia (NH_3), forming a negatively charged chloride ion (Cl^-) and a positively charged ammonium ion (NH_4^+).

Thus, hydrochloric acid (HCl) is a bronsted-lowry acid, and chloride ion (Cl^-) is the conjugate base of the hydrochloric acid (HCl). While ammonia (NH_3) is a bronsted-lowry base, and ammonium ion (NH_4^+) is the conjugate acid of ammonia (NH_3).

The Bronsted-Lowry theory

The Bronsted-Lowry theory (Proton theory of acid and base) is an acid-base reaction theory, introduced by Johannes Nicolaus Bronsted (Danish Chemist) and Thomas Martin Lowry (English Chemist) in 1923. According to the theory, acid and base react with each other and by an exchange of proton acid, forms its conjugate base and the base forms its conjugated acid.

Bronsted-Lowry Theory of Acid and Base

According to Bronsted-Lowry theory, acid is a substance which donates an H+ ion or a proton and forms its conjugate base and the base is a substance which accepts an H+ ion or a proton and forms its conjugate acid.

The Bronsted-Lowry Acids and their Conjugated Bases

The strength of the acid decreases as it descends and the strength of their corresponding conjugate base increases.

Acids	Conjugated base
Perchloric acid ($HCIO_4$)	Perchlorate ion (CIO_4^-)
Hydroiodic acid (HI)	Iodide ion (I^-)
Hydrobromic acid (HBr)	Bromide ion (Br^-)
Hydrochloric acid (HCl)	Chloride ion (Cl^-)
Sulphuric acid ($H2SO_4$)	Hydrogen sulphate ion (HSO_4^-)
Nitric acid (HNO_3)	Nitrate ion (NO_3^-)
Hydronium ion ($H3O^+$)	Water (H_2O)
Hydrogen sulfate ion (HSO_4^-)	Sulfate ion ($SO_4^{2}-$)
Nitrous acid (HNO_4)	Nitrite ion (NO_2^-)
Acetic acid (CH_3COOH)	Acetate ion (CH_3COO^-)
Carbonic acid (H_2CO_3)	Hydrogen carbonate ion (HCO_3^-)
Ammonium ion (NH_4^+)	Ammonia (NH_3)
Hydrogen carbonate ion (HCO_3^-)	Carbonate ion (CO_3^{2-})
Water (H_2O)	Hydroxide ion (OH^-)
Methanol (CH_3OH)	Methoxide ion (CH_3O^-)
Ammonia (NH_3)	Amide ion (NH_2^-)

Lewis Acid

In 1923 G. N. Lewis suggested another way of looking at the reaction between H+ and OH- ions. In the Brnsted model, the OH- ion is the active species in this reaction - it accepts an H+ ion to form a covalent bond. In the Lewis model, the H+ ion is the active speciesit accepts a pair of electrons from the OH- ion to form a covalent bond.

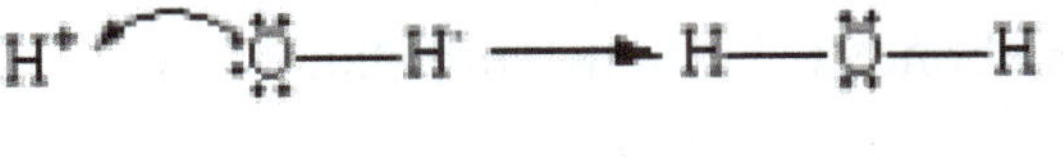

According to Lewis, an acid is a substance having a vacant orbital capable of accepting electron pairs to form a Lewis adduct.

- Lewis Acids are electrophilic.
- Lewis said that acid must not contain a hydrogen ion (H+) or hydroxide ion (OH^-) to be classified as an acid.
- All cations are regarded as lewis acid since they can accept electrons. Examples: Ti^{3+}, Cu^{3+}, Mn^{2+} etc.
- An atom, molecule, or ion with an incomplete valence shell of electrons is regarded as a lewis acid. Examples BF_3, $AlCl_3$.
- Molecules having more than 8 electrons in the central atom are also regarded as lewis acid since they can accept electrons. Examples: $SiBr_4$, SiF_4
- Molecules with multiple bonds between two atoms of different electronegativities are also regarded as lewis acid. Examples: CO_2, SO_2.
- In the Lewis theory of acid-base reactions, bases donate pairs of electrons and acids accept pairs of electrons. A Lewis acid is therefore any substance, such as the H+ ion, that can accept a pair of nonbonding electrons. In other words, a Lewis acid is an electron-pair acceptor. A Lewis base is any substance, such as the OH- ion, that can donate a pair of nonbonding electrons. A Lewis base is therefore an electron-pair donor.
- One advantage of the Lewis theory is the way it complements the model of oxidation-reduction reactions. Oxidation-reduction reactions involve a transfer of electrons from one atom to another, with a net change in the oxidation number of one or more atoms.

- The Lewis theory suggests that acids react with bases to share a pair of electrons, with no change in the oxidation numbers of any atoms. Many chemical reactions can be sorted into one or the other of these classes. Either electrons are transferred from one atom to another, or the atoms come together to share a pair of electrons.
- The principal advantage of the Lewis theory is the way it expands the number of acids and therefore the number of acid-base reactions. In the Lewis theory, an acid is any ion or molecule that can accept a pair of nonbonding valence electrons. In the preceding section, we concluded that Al3+ ions form bonds to six water molecules to give a complex ion.

$$Al^{3+}(aq) + 6\ H_2O(l)\ Al(H2O)_6^{3+}(aq)$$

This is an example of a Lewis acid-base reaction. The Lewis structure of water suggests that this molecule has nonbonding pairs of valence electrons and can therefore act as a Lewis base. The electron configuration of the Al3+ ion suggests that this ion has empty 3s, 3p, and 3d orbitals that can be used to hold pairs of nonbonding electrons donated by neighboring water molecules.

$$Al^{3+} = [Ne]\ 3s^0\ 3p^0\ 3d^0$$

Thus, the Al(H2O)63+ ion is formed when an Al3+ ion acting as a Lewis acid picks up six pairs of electrons from neighboring water molecules acting as Lewis bases to give an acid-base complex, or complex ion.

The Lewis acid-base theroy explains why BF_3 reacts with ammonia. BF_3 is a trigonal planar molecule because electrons can be found in only three places in the valence shell of the boron atom. As a result, the boron atom is sp2 hybridized, which leaves an empty 2pz orbital on the boron atom. BF3 can therefore act as an electron-pair acceptor, or Lewis acid. It can use the empty 2pz orbital to pick up a pair of nonbonding electrons from a Lewis base to form a covalent bond. BF3 therefore reacts with Lewis bases such as NH3 to form acid-base complexes in which all of the atoms have a filled shell of valence electrons, as shown in the figure below:

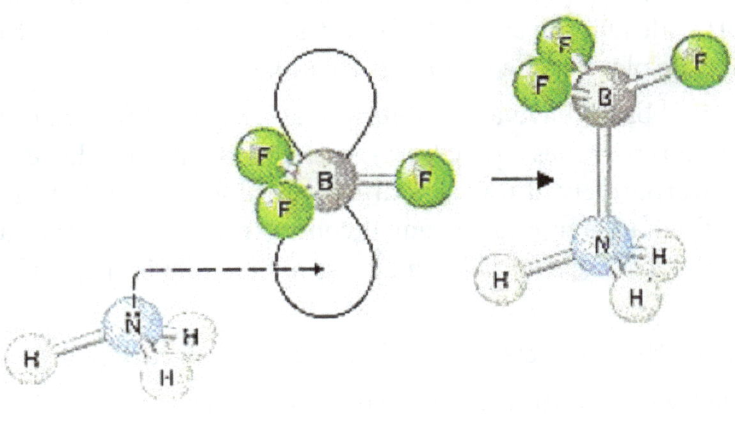

The Lewis acid-base theory can also be used to explain why nonmetal oxides such as CO2 dissolve in water to form acids, such as carbonic acid H_2CO_3.

$$CO_2(g) + H_2O(l)\ H_2CO_3(aq)$$

In the course of this reaction, the water molecule acts as an electron-pair donor, or Lewis base. The electron-pair acceptor is the carbon atom in CO2. When the carbon atom picks up a pair of electrons from the water molecule, it no longer needs to form double bonds with both of the other oxygen atoms as shown in the figure below

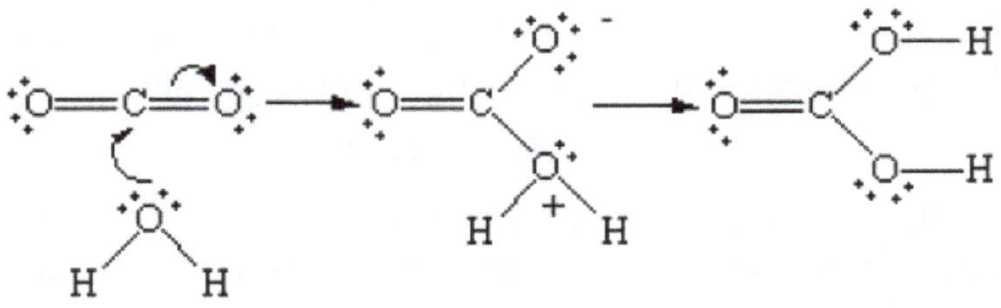

One of the oxygen atoms in the intermediate formed when water is added to CO2 carries a positive charge; another carries a negative charge. After an H+ ion has been transferred from one of these oxygen atoms to the other, all of the oxygen atoms in the compound are electrically neutral. The net result of the reaction between CO2 and water is therefore carbonic acid, H2CO3.

ACID-BASE EQUILIBRIA AND PH CALCULATIONS

Acid–Base Equilibria Acids and bases play a key role in a number of environmentally important chemical reactions, including weathering, transport of metals in solution, and CO2 atmosphere–water equilibria. In this chapter we will develop the concept of an acid and a base, characterize strong and weak acids, develop the pH scale and the concept of buffers, and look at some of the important reactions that take place in the surface environment.

Both acids and bases have an equilibrium state when the rate of forward and backward reactions becomes equal. But this doesn't mean that there is no reaction taking place. Alternatively, it means that the concentration of both reactants and products doesn't change due to the equal concentration of reactant and product. Determination of acid base equilibrium plays a vital role in a specific reaction. It helps in :

- Predicting the products obtained in a specific reaction.
- The relative concentration of the products formed.
- Helps in the identification of the weaker acid and base.
- Weak acids and bases are little common in their nature.

Arrhenius Concept of Acid & Base Dissociation

The Arrhenius concept plays a critical role in explaining the concept of degree of dissociation. As ionization takes place in an aqueous solution, Arrhenius's theory fits perfectly to explain the dissociation process and acid-base equilibria (including multi stage ionization) and ionization constants. It is a way to determine the strength of acids and bases involved in the reaction. However, the strength of particular acid and base depends entirely on the ionization power. The more acid or base gets ionized in the particular reaction, the more is the ionization constant of the reaction. In shorter words, we can say that the acids that have high power to dissolve into their constituent ions are known as strong acids. Thus, they have the capability of donating protons instantly. Hence they are known as good proton donors. On the other hand, strong bases are termed good proton acceptors because they are good at accepting protons.

Strong Acid & Bases

In an aqueous solution, the strong acid is capable of dissolving. It dissociates completely in the aqueous solution resulting in the formation of H+ or hydrogen ions. This concept is based on Arrhenius's theory. Similarly, when the strong base is made to dissolve into the aqueous solution, it completely dissolves into OH- ions.

In short terms, an acid that has the power to dissolve 100% in the aqueous solution, is known as a very very strong acid. And the base that has the power to dissolve 100 % in the aqueous medium, they are termed as a very very strong base. Let's understand them by taking a specific example.

Let's first talk about a very very strong acid, HCl.

$$HCl + H_2O \rightarrow H3O+ + Cl-$$

When HCl is dissolved in an aqueous solution, water, the acid dissolves to form H3O+ ions. The acid completely dissolves which indicates that the hydrochloric acid is quite strong.

Let's now move to the example of a very very strong base, NaOH.

$$NaOH + H2O \rightarrow Na + + OH-$$

In this reaction, the aqueous medium is mixed with NaOH, it dissolves completely into sodium ions and OH- ions.

A few examples of strong acids are Hydrochloric Acid, Hydrobromic Acid, Hydroiodic Acid, Nitric Acid, Perchloric Acid, Sulfuric Acid, etc.

A few examples of strong bases are Barium Hydroxide, Calcium Hydroxide, Potassium Hydroxide, Sodium Hydroxide, etc.

Weak Acids & Bases

Alternative to strong acids and bases, the weak acids and bases don't completely dissolve in the aqueous medium. Let's have a look at the general format of dissociation of a weak acid.

$$HA + H2O \Rightarrow H3O+ + A$$

In such a reaction, the arrow is double-sided. This indicates that the reaction is capable of either moving forward to form the ions or it can move to the backward direction to increase the concentration of the reactant. This type of reaction is also known as a recombining reaction.

Equilibrium constant acid-base

Also known as ionization constants, or dissociation constants, acids, and bases have their equilibrium constants denoted by Ka and Kb respectively. The equilibrium constant acid-base is known to determine the relative strength of the chemical compounds. It's important to determine the equilibrium constant to know the rate of dissociation and the extent of dissociation of chemical compounds. If you know the equilibrium constant of a particular reaction, it's easy to calculate the ionization percentage along with the pH value of a reaction.

Proton Transfer

In a reaction, when the proton gets transferred from acid to base, the pair formed is known as conjugate acid-base pairs.

$$HBr + H2O \rightarrow Br- + H3O+$$

In the above reaction, we have a strong acid, HBr, and a strong base, H2O. When the reaction occurs, it leads to the formation of Brand H3O+ ions. Meaning, the proton transfer takes place from Hydrobromic Acid to water. Thus, in this reaction, water and hydronium ions are referred to as conjugate acid-base pairs. One thing to note is that water can act as both acid and base, depending on the reaction and the conditions. In other words, water has an amphoteric nature (the compound that can act as acid and base).

Equilibrium constant acid-base reaction

Acids:-

The equilibrium constant in an acidic reaction can be written in the form –

$$Ka = [H_3O+]\,[A-]\,/[HA],$$

Where $[H_3O+]$ is the concentration of hydronium ions

[A-] is the concentration of the conjugate base

[HA] is the concentration of acids

Bases:-

To define the equilibrium constant of the base here is the simple formula

$$Kb = [BH+]\,[OH-]\,/[B],$$

Where [BH+] is the concentration of conjugate acids

[OH-] is the concentration of Hydroxide ion

[B] is the concentration of the base.

pH Calculations

We are surrounded by dilute solutions of acids and bases, inside and out. The orange juice and coffee that help us start our day are acidic, and the gastric juices they mix with in our stomach are also naturally acidic. Many of our household cleaning liquids, like the detergent solution in our washing machines and the bleach we add to whiten our clothes, are basic. Now that you know about reversible reactions and how solutions are described in terms of molarity, you will be able to understand the origin of the pH scale for describing acids and what the pH value says about an acidic or basic solution.

According to the Arrhenius theory of acids and bases, when an acid is added to water, it donates an H^+ ion to water to form H3O+ (often represented by H^+). The higher the concentration of H_3O^+ (or H+) in a solution, the more acidic the solution is. An Arrhenius base is a substance that generates hydroxide ions, OH^-, in water. The higher the concentration of OH- in a solution, the more basic the solution is.

Pure water undergoes a reversible reaction in which both H+ and OH- are generated.

$$H_2O(l)\ H^+(aq) + OH^-(aq)$$

The equilibrium constant for this reaction, called the water dissociation constant, K_w, is 1.01 × 10-14 at 25 °C.

$$K_w = [H+][OH-] = 1.01 \times 10-14 \text{ at } 25\ °C$$

Because every H+ (H3O+) ion that forms is accompanied by the formation of an OH ion, the concentrations of these ions in pure water are the same and can be calculated from K_w.

$$K_w = [H^+][OH^-] = (x)(x) = 1.01 \times 10^{-14}$$

$$x = [H^+] = [OH-] = 1.01 \times 10^{-7}\ M$$

(1.005 × 10-7 M before rounding)

The equilibrium constant expression shows that the concentrations of H+ and OH in water are linked. As one increases, the other must decrease to keep the product of the concentrations equal to 1.01 × 10-14 (at 25 °C). If an acid, like hydrochloric acid, is added to water, the concentration of the H+ goes up, and the concentration of the OH goes down, but the product of those concentrations remains the same. An acidic solution can be defined as a solution in which the [H+] > [OH-]. The example below illustrates this relationship between the concentrations of H+ and OH- in an acidic solution.

Concentrations of H+ and OH- in Dilute Acid and Base Solutions at 25 °C

[H+]	[OH-]
1.0 M	1.0×10^{-14} M
1.0×10^{-3} M	1.0×10^{-11} M
1.0×10^{-7} M	1.0×10^{-7} M
1.0×10^{-10} M	1.0×10^{-4} M
1.0×10^{-14} M	1.0 M

We could describe the relative strengths of dilute solutions of acids and bases by listing the molarity of H^+ for acidic solutions and the molarity of OH^- for basic solutions. There are two reasons why we use the pH scale instead. The first reason is that instead of describing acidic solutions with $[H^+]$ and basic solutions with $[OH^-]$, chemists prefer to have one scale for describing both acidic and basic solutions. Because the product of the H^+ and OH^- concentrations in such solutions is always 1.01×10^{-14} at 25 °C, when we give the concentration of H+, we are indirectly also giving the concentration of OH^-. For example, when we say that the concentration of H^+ in an acidic solution at 25 °C is 10^{-3} M, we are indirectly saying that the concentration of OH^- in this same solution is 10^{-11} M. When we say that the concentration of H+ in a basic solution at 25 °C is 10^{-10} M, we are indirectly saying that the OH^- concentration is 10^{-4} M. The pH concept makes use of this relationship to describe both dilute acid and dilute base solutions on a single scale.

The next reason for using the pH scale instead of H^+ and OH^- concentrations is that in dilute solutions, the concentration of H^+ is small, leading to the inconvenience of measurements with many decimal places, such as 0.000001 M H^+, or to the potential confusion associated with scientific notation, as with 1×10^{-6} M H^+. In order to avoid such inconvenience and possible confusion, pH is defined as the negative logarithm of the H+ concentration.

$$pH = -\log[H^+]$$

Instead of saying that a solution is 0.0000010 M H+ (or 1.0×10^{-6} M H^+) and 0.000000010 M OH- (or 1.0×10^{-8} M OH^-), we can indirectly convey the same information by saying that the pH is 6.00.

$$pH = -\log[H^+] = -\log(1.0 \times 10^{-6}) = 6.00$$

When taking the logarithm of a number, report the same number of decimal positions in the answer as you had significant figures in the original value. Because 1.0 × 10-6 has two significant figures, we report 6.00 as the pH for a solution with 1.0 × 10-6 M H+. The table below shows a range of pH values for dilute solutions of acid and base.

pH of Dilute Solutions of Acids and Bases at 25 °C

$[H^+]$	$[OH^-]$	pH
1.0	1.0×10^{-14}	0.00
1.0×10^{-1}	1.0×10^{-13}	1.00
1.0×10^{-2}	1.0×10^{-12}	2.00
1.0×10^{-3}	1.0×10^{-11}	3.00
1.0×10^{-4}	1.0×10^{-10}	4.00
1.0×10^{-5}	1.0×10^{-9}	5.00
1.0×10^{-6}	1.0×10^{-8}	6.00
1.0×10^{-7}	1.0×10^{-7}	7.00
1.0×10^{-8}	1.0×10^{-6}	8.00
1.0×10^{-9}	1.0×10^{-5}	9.00
1.0×10^{-10}	1.0×10^{-4}	10.00
1.0×10^{-11}	1.0×10^{-3}	11.00
1.0×10^{-12}	1.0×10^{-2}	12.00
1.0×10^{-13}	1.0×10^{-1}	13.00
1.0×10^{-14}	1.0	14.00

The above table illustrates several important points about pH. Notice that

- When the solution is acidic ($[H^+] > [OH^-]$), the pH is less than 7. • When the solution is basic ([OH-] > [H+]), the pH is greater than 7.
- When the solution is neutral ([H+] = [OH-]), the pH is 7. (Solutions with pH's between 6 and 8 are often considered essentially neutral.)

Also notice that

- As a solution gets more acidic (as [H+] increases), the pH decreases. • As a solution gets more basic (higher [OH-]), the pH increases.
- As the pH of a solution decreases by one pH unit, the concentration of H+ increases by ten times.
- As the pH of a solution increases by one pH unit, the concentration of OH increases by ten times.

The pH, [H+], and [OH-] of some common solutions are listed in the figure below.

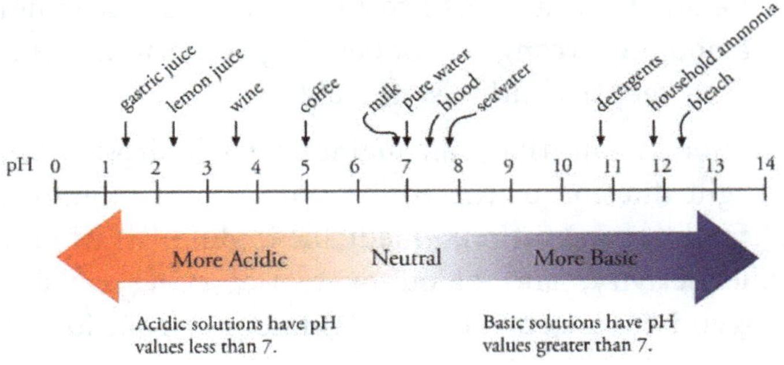

Notice that gastric juice in our stomach has a pH of about 1.4, and orange juice has a pH of about 2.8. Thus, gastric juice is more than ten times more concentrated in H+ than orange juice. The pH difference of about 4 between household ammonia solutions (pH about 11.9) and milk (pH about 6.9) shows that household ammonia has about ten thousand (104) times the hydroxide concentration of milk.

pH of Common Substances

Acidic solutions have pH values less than 7, and basic solutions have pH values greater than 7. The more acidic the solution is, the lower its pH. The more basic a solution is, the higher the pH. The corresponding H+ and OH- concentrations are shown in units of molarity. Notice that a decrease of one pH unit corresponds to a ten-fold increase in [H+], and an increase of one pH unit for a basic solution corresponds to a ten-fold increase in [OH-].

BUFFER SOLUTIONS AND THEIR SIGNIFICANCE

Buffer Solution is a special aqueous solution that resists the change in its pH when some quantity of acid and Base is added. Many fluids, such as blood, have specific pH values of 7.14, and variations in these values indicate that the body is malfunctioning. The change in pH of Buffer Solutions on adding a small quantity of acid or bases is very minimal and hence they are used to make solutions that resist the change in pH.

Buffer Solution Definition

A buffer solution is a solution that resists changes in hydrogen ion concentration when a modest amount of acid or base is added. In other words, such solutions are known to have reverse acidity and reverse basicity and to keep a reasonably steady pH value. A good example of a natural buffer solution is human blood. Despite eating a wide array of meals, our blood maintains a pH of roughly 7.35.

Buffer Solution is a water-based solvent-based solution made up of a weak acid and its conjugate base, or a weak base and its conjugate acid. They are resistant to changes in pH caused by dilution or the addition of relatively small amounts of acid or alkali. When a small amount of strong acid or strong base is added, the pH of the buffer solution changes very little. As a result, they're used to maintaining a steady pH. When a tiny amount of strong acid or base is given to it, its pH varies very little, and it is thus used to keep a solution's pH stable.

A buffer solution is one that can maintain its hydrogen ion concentration (pH) with just slight dilution or the addition of a small amount of acid or base. Fermentation, food preservation, medicine administration, electroplating, printing, enzyme activity, and blood oxygen-carrying capability all require particular hydrogen ion concentrations (pH) in buffer solutions. Buffer solutions are made up of a weak acid and its conjugate base or a weak base and its conjugate acid that can maintain pH.

Types of Buffer Solutions

Acidic and basic buffers are the two types of buffer solutions that are extensively classified. These are discussed in greater depth down below.

Acidic Buffers

Acidic buffer solutions are made up of equimolar amounts of a weak acid and its salt, as well as a strong base. These solutions are used to keep the environment acidic. An acid buffer is made by combining a weak acid and its salt with a strong base to create an acidic pH.

The pH of an aqueous solution containing an equal amount of acetic acid and sodium acetate is 4.74. Furthermore, the pH of these liquids is less than seven. These solutions are made up of a weak acid and its salt.

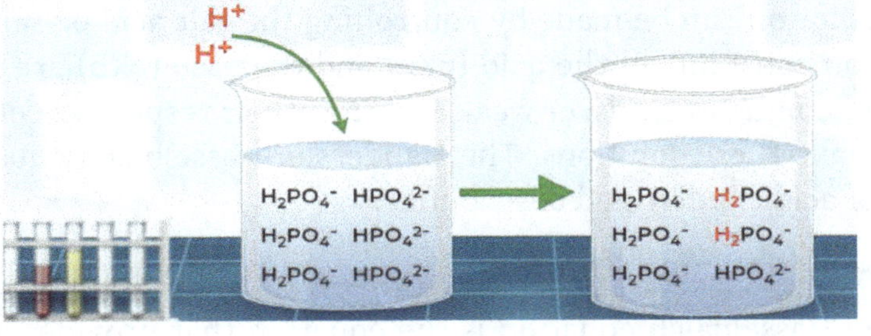

Basic or Alkaline Buffers

A weak base and its salt are equimolar with a strong acid in a basic buffer solution. These buffer solutions are employed to keep basic conditions. To generate a basic buffer with a basic pH, a weak base and its salt are combined with a strong acid.

An aqueous solution of equal parts of ammonium hydroxide and ammonium chloride has a pH of 9.25. These solutions have a pH of greater than seven. They contain a weak base and a weak base salt.

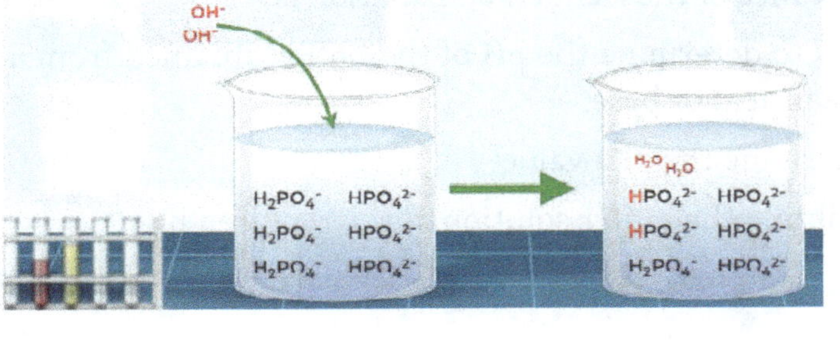

Mechanism of Buffering Action

To understand how a buffer works, think about the example of a buffer solution created by combining sodium acetate and acetic acid. Acetate acid, as implied by its name, is an acid with the chemical formula CH3COOH, whereas sodium acetate dissociates in solution to produce the conjugate base CH3COO– acetate ions. The reaction equation is:

$$CH_3COOH\ (aq) + OH^-\ (aq) \rightarrow CH_3COO^-\ (aq) + H_2O\ (aq)$$

Now, the acetate ion can be neutralized by adding a strong acid to this solution as,

$$CH_3COO^-\ (aq) + H+\ (aq) \rightarrow CH_3COOH\ (aq)$$

Thus, the original buffer reaction equilibrium changes, and hence, the pH remains constant.

Preparation of Buffer Solution

A buffer solution can be made by controlling the salt acid or salt base ratio if the dissociation constants of the acid (pKa) and the base (pKb) are known. Weak bases and their conjugate acids, or weak acids and their respective conjugate bases, are used to make these solutions. The Handerson-Hasselbalch equation and the preparation of acidic buffer and basic buffer.

Henderson-Hasselbalch Equation

Henderson-Hasselbalch equation is the equation that provides a relationship between the pH of acids and their pKa.

The Henderson-Hasselbalch equation is,

$$pH = pKa + \log10\ ([A-] / [HA])$$

where

$[A^-]$ denotes the molar concentration of the Conjugate Base (of the Acid)

[HA] denotes the molar concentration of the Weak Acid

Importance of Henderson Equation

The significance of the Henderson Equation is,

- It helps to determine the pH of the buffer obtained from a mixture of salt and weak acid or base.
- To determine the pKa value.
- To maintain the buffer solution of the required pH.

Limitations of Henderson-Hasselbalch Equation

The Henderson-Hasselbalch Equation is not applicable in many cases are:

- This equation does not explain the water self-dissociation tendency.
- The assumptions stated by this equation are not applicable when strong acids or bases are associated.
- A buffer solution can easily be prepared using Henderson-Hasselbalch Equation

Preparation of Acid Buffer

In an acid buffer solution with a strong base (KOH), consider a weak acid (HA) and its salt (KA). The weak acid (HA) ionizes, and the equilibrium is as follows:

$$H_2O + HA \text{ Ë! } H^+ + A^-$$

The acid dissociation constant is,

$$K_a = ([H^+] [A^-])/[HA]$$

RHS and LHS take negative logs:

$$-\log K_a = -\log [H^+] - \log ([A^-]/[HA])$$

$$pK_a = pH - \log ([salt]/[acid])$$

$$pH = pK_a + \log ([salt]/[acid])$$

pH of acid buffer solution = pK_a + ([salt]/[acid])

Preparation of Base Buffer

Consider a basic buffer solution with strong acid, salt (BA), and a weak base (B). As a result, the basic buffer solution will be,

pOH = pKb + log ([salt]/[base])

pOH of a basic buffer solution = pKb + log ([salt]/[acid])

pH of a basic buffer solution = pKa – log ([salt]/[acid])

Buffering Capacity

Buffering Capacity is the property of solutions which is represented by the symbol b. It is defined as the ratio of the moles of acid or base required to change the pH of the solution by 1 and the pH change and the volume of buffers.

b = millimoles /(DpH)

Uses of Buffer Solutions

Various uses of Buffer solution are,

- Buffer solutions are referred to by several different names, including pH buffers and hydrogen ion buffers.
- Many organisms employ buffer solutions to maintain an optimum pH for enzyme activity.
- The use of bicarbonate and carbonic acid buffer system to manage the pH of animal blood is an example of a buffer used in pH regulation.
- Enzyme function may be hindered, features may be lost, or the enzymes may even denature if certain buffers are not present. The enzymes' catalytic function can be completely deactivated by this denaturation process.

pH Maintenance

How the pH of any solution is kept constant can easily be explained with the help of the example given below,

Take formic acid and sodium formate ions in an aqueous solution they behave as buffer solution and their equilibrium reaction is,

- $HCOOH \Rightarrow H^+ + HCOO^-$
- $HCOONa \Rightarrow Na^+ + HCOO^-$

If the strong acids are added to the above buffer solution, the hydrogen ion (H+ ions) combine with the HCOO– ions to give a weakly ionized formic acid, and thus, the pH of the solution is kept constant or slightly changed.

If the strong base is added to the above buffer solution, the hydroxide ion (OH– ions) combine with the H+ ions available in the solution to give water molecules, and thus, the pH of the solution is kept constant. The reaction occurring in the process is given below.

$$HCOOH + OH^- \Rightarrow HCOO^- + H_2O$$

Thus, the pH of the buffer solution is kept constant.

ACID-BASE CATALYSIS

Acid and bases are the most widely used reagents in chemical laboratories. In various reactions, it performs the role of catalyst. Let us discuss what acid-base catalysis reaction is? Acid-base catalysis is the addition of an acid or a base to a chemical reaction to speed up the reaction without consuming the acid or base. Acid catalysis, as in the decomposition of sucrose into glucose and fructose in sulfuric acid, or base catalysis, as in the addition of hydrogen cyanide to aldehydes and ketones in the presence of sodium hydroxide, are two examples of catalytic reactions. Acids and bases both catalyze a variety of reactions.

General Acid-Base Catalysis Mechanism

The Bronsted-Lowry principle of acids and bases describes the mechanism of acid and base-catalyzed reactions as an initial transfer of protons from an acidic catalyst to the reactant or from the reactant to a basic catalyst. The reaction involves the exchange of an electron pair donated by a base catalyst or embraced by an acid catalyst, according to the Lewis theory of acids and bases.

Acid catalysis is used in many industrial processes, including the conversion of petroleum hydrocarbons to gasoline and other products. High-molecular-weight hydrocarbons are decomposed (cracked) using alumina-silica catalysts (Bronsted-Lowry acids), unsaturated hydrocarbons are polymerized using sulfuric acid or hydrogen fluoride (Bronsted-Lowry acids), and aliphatic hydrocarbons are isomerized using aluminium chloride (Bronsted-Lowry acids) (a Lewis acid). There are two types of acid catalysis: basic acid catalysis and general acid catalysis.

Characteristics of Acid-base Catalysis

The great majority of enzyme reactions–the chemical reactions on which life depends– take place in water, under "physiological" conditions, near pH 7. Yet this is where chemical reactions are normally at their slowest. Reactive compounds are readily hydrolyzed by adding strong acid or base in vitro. But this is not a practical proposition for an enzyme: proteins are denatured in the presence of strong acid or base. Nevertheless, the way we think about catalysis by enzymes is based on our understanding of acid–base catalysis in vitro.

The acid-base catalysis is illustrated by the pH–rate profiles shown in Fig. 6.1: the three plots are representative of the behavior of most types of substrate undergoing most sorts of reaction in water. The lower curve (I) represents the reaction in vitro of a typical, unreactive compound: it shows only acid- and base-catalyzed reactions, and reaction is very slow at the minimum, near pH 7. For the most reactive compounds, an additional feature (curve II) is a pH-independent region, where the uncatalyzed reaction with water becomes faster than the acid- and base-catalyzed reactions near neutrality. Finally, curve III is the pH–rate profile for a typical enzyme-catalyzed reaction: it is much faster than the others (signified by the break in the ordinate), and it is also qualitatively quite different, since now the rate reaches a maximum near pH 7. Enzymes are 'designed' to operate near pH 7, and typically show pH optima in this region, with rates falling off at higher and lower pH values.

The acid- and base-catalyzed reactions at high and low pH are not directly relevant to catalysis by enzymes. They typically involve activation of the substrate by the addition or removal of a proton in a rapid pre-equilibrium, followed by the rate-determining reaction of the conjugate acid or base.

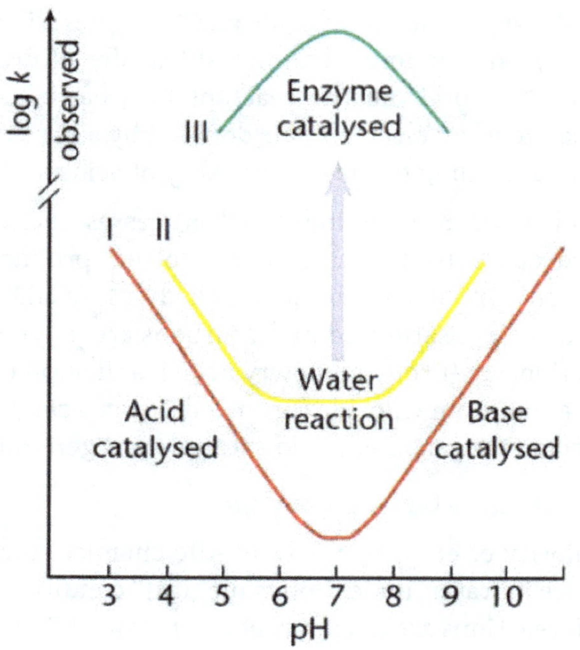

Fig. 6.1 Specific acid–base catalysis and enzyme catalysis compared. (Kirby A J. 2001)

7

REDOX REACTIONS AND ELECTROCHEMISTRY

REDOX REACTIONS

Redox reactions are oxidation-reduction chemical reactions in which the reactants undergo a change in their oxidation states. The term 'redox' is a short form of reduction oxidation. All the redox reactions can be broken down into two different processes: a reduction process and an oxidation process.

A redox reaction can be defined as a chemical reaction in which electrons are transferred between two reactants participating in it. This transfer of electrons can be identified by observing the changes in the oxidation states of the reacting species.

An illustration (Fig. 7.1) detailing the electron transfer between two reactants in a redox reaction is provided below.

In the illustration provided below, it can be observed that the reactant, an electron, was removed from reactant A, and this reactant is oxidized. Similarly, reactant B was handed an electron and was, therefore, reduced.

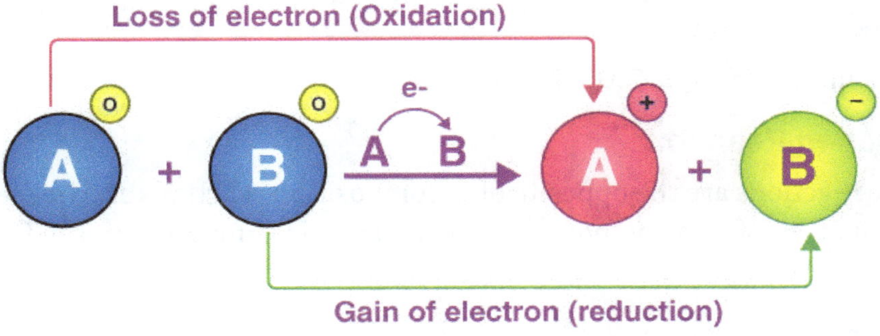

Fig. 7.1 The electron transfer between two reactants in a redox reaction

The loss of electrons and the corresponding increase in the oxidation state of a given reactant is called oxidation. The gain of electrons and the corresponding decrease in the oxidation state of a reactant is called reduction.

Electron-accepting species which tend to undergo a reduction in redox reactions are called oxidizing agents. An electron-donating species which tends to hand over electrons can be referred to as a reducing agent. These species tend to undergo oxidation. It can be noted that any redox reaction can be broken down into two half-reactions, namely, the oxidation half-reaction and the reduction half-reaction.

When writing these half-reactions separately, each of them must be balanced in a way that all the electrons are accounted for.

Types of Redox Reactions

The different types of redox reactions are

- Decomposition Reaction
- Combination Reaction
- Displacement Reaction
- Disproportionation Reactions

Decomposition Reaction

This kind of reaction involves the breakdown of a compound into different compounds. Examples of these types of reactions are

- $2NaH \rightarrow 2Na + H_2$
- $2H_2O \rightarrow 2H_2 + O_2$
- $Na_2CO_3 \rightarrow Na_2O + CO_2$

All the above reactions result in the breakdown of smaller chemical compounds in the form of $AB \rightarrow A + B$

But, there is a special case that confirms that all the decomposition reactions are not redox reactions.

For example, $CaCO_3 \rightarrow CaO + CO_2$

Combination Reaction

These reactions are the opposite of decomposition reactions and hence, involve the combination of two compounds to form a single compound in the form of A + B ® AB.

For example,

- $H_2 + Cl_2 \rightarrow 2HCl C + O_2 \rightarrow CO_2$
- $4Fe + 3O_2 \rightarrow 2Fe_2O_3$

Displacement Reaction

In this kind of reaction, an atom or an ion in a compound is replaced by an atom or an ion of another element. It can be represented in the form of X + YZ ® XZ + Y.

Further displacement reactions can be categorised into

- Metal displacement reaction
- Non-metal displacement reaction

Metal Displacement

In this type of reaction, a metal present in the compound is displaced by another metal. These types of reactions find their application in metallurgical processes where pure metals are obtained from their ores.

For example, $CuSO_4 + Zn \rightarrow Cu + ZnSO_4$

Non-metal Displacement

In this type of reaction, we can find a hydrogen displacement and, sometimes, rarely occurring reactions involving oxygen displacement.

Disproportionation Reactions

Disproportionation reactions are known as reactions in which a single reactant is oxidized and reduced.

For example, $P_4 + 3NaOH + 3H_2O \rightarrow 3NaH_2PO_2 + PH_3$

Examples of Redox Reactions

A few examples of redox reactions, along with their oxidation and reduction half reactions, are provided in this subsection.

Example 1: Reaction between Hydrogen and Fluorine

In the reaction between hydrogen and fluorine, the hydrogen is oxidized, whereas the fluorine is reduced. The reaction can be written as follows.

$$H_2 + F_2 \rightarrow 2HF$$

The oxidation half-reaction is: $H_2 \rightarrow 2H^+ + 2e^-$

The reduction half-reaction is: $F_2 + 2e^- \rightarrow 2F-$

The hydrogen and fluorine ions go on to combine in order to form hydrogen fluoride.

Example 2: Reaction between Zinc and Copper

This is a type of metal displacement reaction in which copper metal is obtained when zinc displaces the Cu2+ion in the copper sulphate solution, as shown in the reaction below:

$$Zn\ (s) + CuSO_4\ (aq) \rightarrow ZnSO_4\ (aq) + Cu\ (s)$$

The oxidation half-reaction can be written as $Zn \rightarrow Zn^{2+} + 2e^-$

The reduction half-reaction can be written as $Cu^{2+} + 2e^-$ ® Cu

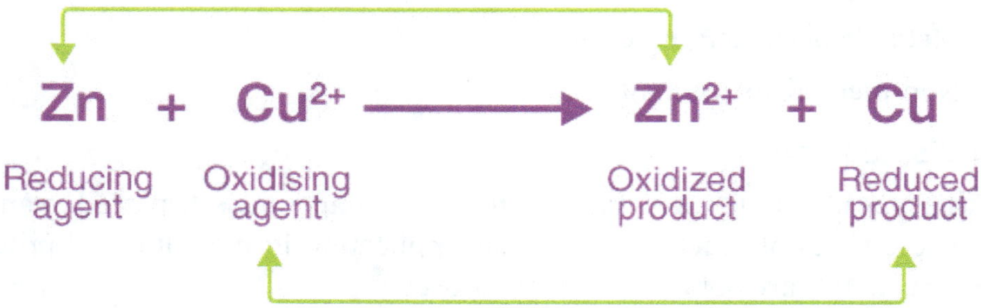

Fig. 7.2

Thus, copper is displaced from the copper sulphate solution by zinc in a redox reaction.

Example 3: Reaction between Iron and Hydrogen Peroxide

Fe2+ is oxidized to Fe3+ by hydrogen peroxide when an acid is present. This reaction is provided below.

$2Fe^{2+} + H_2O_2 + 2H^+ \rightarrow 2Fe^{3+} + 2H_2O$

Oxidation half-reaction: $Fe^{2+} \rightarrow Fe^{3+} + e^-$

Reduction half-reaction: $H_2O_2 + 2e^- \rightarrow 2\ OH^-$

Thus, the hydroxide ion formed from the reduction of hydrogen peroxide combines with the proton donated by the acidic medium to form water.

Oxidation and Reduction Reaction

In order to understand redox reactions, let us first deal with oxidation and reduction reactions individually.

What Is the Oxidation Reaction?

Oxidation may be defined as the loss of electrons from a substance; the other definition of oxidation reactions states that the addition of oxygen or the more electronegative element or removal of hydrogen or the more electropositive element from a substance is called an oxidation reaction.

Following are some examples of oxidation reactions:

$$2S(s) + O_2 (g) \rightarrow SO_2 (g)\ CH_4 (g) + 2O_2 (g) \rightarrow CO_2 (g) + 2H_2O (l)$$

Electrochemistry

- Production of chemical change by electrical energy, i.e., the phenomenon of electrolysis
- Conversion of chemical energy into electrical energy, i.e., the generation of electricity by spontaneous redox reactions.

Electricity can be produced when electrons move from one element to another in certain types of reactions (such as redox reactions). Typically, electrochemistry deals with the overall reactions when multiple redox reactions occur simultaneously, connected via some external electric current and a suitable electrolyte. In other words, electrochemistry

is also concerned with chemical phenomena that involve charge separation (as seen commonly in liquids such as solutions). The dissociation of charge often involves charge transfer that occurs homogeneously or heterogeneously between different chemical species.

Electrochemical Cell

A spontaneous chemical process is one which can take place on its own, and in such a process, the Gibbs free energy of a system decreases. In electrochemistry, spontaneous reaction (redox reaction) results in the conversion of chemical energy into electrical energy. The reverse process is also possible where a non-spontaneous chemical reaction occurs by supplying electricity. These interconversions are carried out in equipment called an electrochemical cell.

Types of Electrochemical Cell

Electrochemical cells are of two types: galvanic cells and electrolytic cells

Galvanic Cell

The galvanic cell converts chemical energy into electrical energy, i.e., electricity can be obtained with the help of a redox reaction. The oxidation and reduction take place in two separate compartments. Each compartment consists of an electrolyte solution and a metallic conductor, which acts as an electrode. The compartment containing the electrode and the solution of the electrolyte is called half cells.

For example, the Daniell cell is a galvanic cell in which zinc and copper are used for a redox reaction to take place.

$$Zn(s) + Cu^{+2}(aq) \rightarrow Zn^{+2} + Cu(s)$$

At anode (oxidation half),

$$Zn(s) \rightarrow Zn^{+2} + 2e^-$$

At cathode(reduction half),

$$Cu^{+2}(aq) + 2e^- \rightarrow Cu(s)$$

Salt bridge: The salt bridge (Fig. 7.3) is usually an inverted U-tube filled with a concentrated solution of inert electrolytes. It is used to maintain the charge balance and to complete the circuit by allowing the flow of ions through it. It contains a gel in which inert electrolytes like KNO3 or K2SO4 are mixed. Through the salt bridge, negative ion flows towards the anode and positive ion flows to the cathode, and the charge balance is maintained, and the cell keeps on functioning.

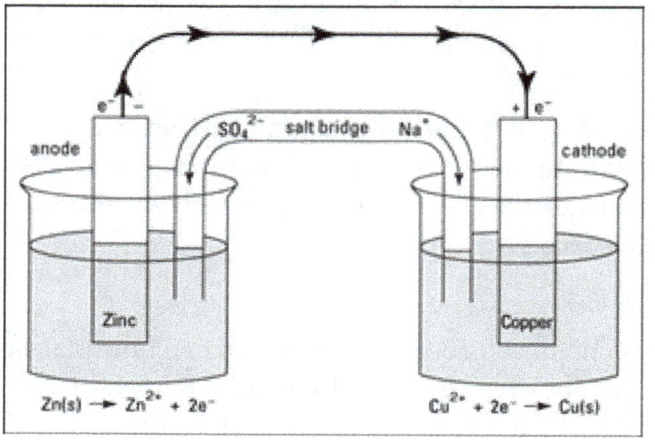

Fig. 7.3 Salt bridge

Electrode potential: In a galvanic cell, when two-electrode are dipped in their respective ion, there is a tendency for one of the electrodes (anode) to undergo oxidation, whereas the ion of the other electrode (cathode) has the tendency to gain an electron. This tendency of losing of electrons (oxidation) or gaining of electrons (reduction) is called electrode potential.

Standard electrode potential (E^0): Standard electrode potential is defined as the electrode potential of an electrode relative to a standard hydrogen electrode under standard conditions. The standard conditions taken are as follows:

- 1 molar concentration of each ion in the solution.
- A temperature of 298 K.
- 1 bar pressure.

Electrochemical Series

A series of the standard electrode has been established by measuring the potential of various electrodes versus standard hydrogen electrodes (SHE). When the electrodes (metals and non-metals) in contact with their ions are arranged on the basis of the values of their standard reduction potential or standard oxidation potentials, the resulting series is called the electrochemical series.

Cell potential or emf of a cell: In the galvanic cell, there are two half-cells, the oxidation half-cell (anode) and the reduction half-cell (cathode). Due to the difference in the potentials of these half-cells, the electric current moves from the electrode of higher potential (cathode) to the electrode of lower potential (anode). The difference between the electrode potential of the two half-cells is called cell potential or emf of a cell.

$E^o_{cell} = E_{Cathode} - E_{anode}$

$E_{Cathode}$ = standard reduction potential of the cathode.

E_{anode} = standard reduction potential of the anode.

If E^o_{cell} is positive, then the reaction is feasible.

If E^o_{cell} is negative, then the reaction is not feasible.

Electrode and cell potentials - Nernst equation: The electrode potential and the emf of the cell depend upon the nature of the electrode, temperature and the activities (concentrations) of the ions in solution.

For the general electrochemical reaction of the type:

$aA + bB \rightarrow cC + dD$

$E_{cell} = E^o_{cell} - RT/ nF \ln [C]^c [D]^d / [A]^a [B]^b$

$E_{cell} = E^o_{cell} - 2.0303\ RT / nF \log [C]^c [D]^d / [A]^a [B]^b$

$E_{cell} = E^o_{cell} - 0.0591 / n \log [C]^c [D]d / [A]^a [B]^b$ at 298K

Electrolytic Cell

The electrolytic cell converts electrical energy to chemical energy. Here, the electrodes are dipped in an electrolytic solution containing cations and anions. On supplying current, the ions move towards electrodes of opposite polarity, and simultaneous reduction and oxidation take place.

For example, in the electrolysis of molten sodium chloride, sodium chloride is melted (above 801oC), two electrodes are inserted into the melt, and an electric current is passed through the molten salt. The chemical reaction that takes place at the electrodes is as follows:

- Sodium-ion migrates to the cathode, where sodium ion gains one electron and reduces to sodium metal.

$$Na^+ + e^- \rightarrow Na$$

- Chloride ions migrate towards the anode, where it loses one electron and gets oxidised to chlorine gas.

$$Cl^- \rightarrow 1/2\ Cl_2 + e^-$$

The overall reaction is the breakdown of sodium chloride into its elements

$$2NaCl \rightarrow 2Na(s) + Cl_2(g)$$

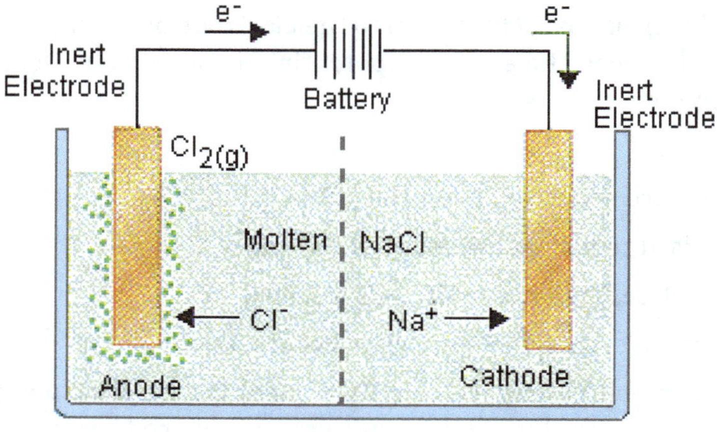

Fig. 7.4

Preferential discharge of ions: When there is more than one cation or anion, the process of discharge becomes competitive in nature. For example, in the electrolysis of NaCl solution, apart from Na+ and Cl–ions, the solution of sodium chloride also contains H+ and OH– ions due to the ionisation of water. When the potential difference is applied between the two electrodes, Na+ and H+ ions move towards the cathode and Cl– and OH– ions move towards the anode. At the cathode, H+ ions get reduced in preference to giving hydrogen gas because hydrogen has a higher reduction potential than sodium. Similarly, at the anode, Cl– ions are oxidised in preference to OH– to give chlorine gas.

Faraday's Law of Electrolysis

The relationship between the quantity of electric charge passed through an electrolyte, and the amount of substance deposited at the electrodes was given by Faraday in 1834, in the form of the law of electrolysis.

Faraday's First Law

When an electric current is passed through an electrolyte, the amount of substance deposited is proportional to the quantity of electric charge passed through the electrolyte.

If W is the mass of the substance deposited by passing Q coulomb of charge, then according to this law:

$$W \Rightarrow Q$$

Now, $Q = I \times t$

$$W \Rightarrow I \times t$$

$$W = z \times I \times t$$

Where z is a constant, known as electrochemical equivalent and is characteristic of a substance deposited.

Faraday's constant (F) – It is the charge possessed by 1 mole of electrons, and it is equal to 96500 coulombs (approx.). In terms of Faraday's constant, the number of gram equivalent of electrolyte discharged at an electrode is equal to Faraday's passed,

$$W = E \times Q / 96500$$

Faraday's Second Law

When the same quantity of charge is passed through different electrolytes, then the mass of different substances deposited at the respective electrodes will be in the ratio of their equivalent masses.

Mathematically, it is represented as,

$$W_1 / W_2 = Z_1 / Z_2$$

Where, W1 and W2 are the weight of two substances which are deposited at their respective electrodes, and Z_1 and Z_2 are their respective equivalent weight.

OXIDATION-REDUCTION DEFINITIONS AND HALF-REACTIONS

Oxidation and Reduction reactions are simply called Redox reactions. There are chemical reactions in which the oxidation number of the chemical species involved in the reaction changes. Oxidation and reduction Reactions involve a wide variety of processes. For example, oxidation-reduction reactions are so common that they can be the breathing process, burning of wood, rusting, fruit ripening, photosynthesis, and many more. Generally, these redox reactions involve the transfer of oxygen or hydrogen atoms.

What is Reduction?

Reduction is a process that involves the addition of hydrogen or any other electropositive element or the removal of oxygen or any other electronegative element, according to classical or older concepts.

Process through which an atom or an ion gains one or more electrons is known as reduction, according to the electronic concept.

Oxidant

Oxidant, also known as an oxidizing agent, is a substance that can accept electrons.

Examples of Oxidants

In periodic table, molecules of electronegative elements (Oxygen) and halogens (Fluorine, Chlorine, Bromine, and Iodine) have high oxidation states. The following compounds have the highest Oxidation state, e.g: K2Cr2O7, KMnO4, HClO4, H2SO4, and KClO4.

Reductant

A reductant, which is also known as a reducing agent is a chemical substance that can donate electrons.

Examples of Reductants

Metals such as sodium, Aluminum, and Zinc and non-metals like Calcium, Phosphorous, Sodium and Hydrogen gas are strong reducing agents.

Like oxidation reactions, reduction reactions are defined as the gain of electrons. Any substance that gains electrons during a chemical reaction gets reduced.

In other forms, the reduction reaction is stated as the addition of hydrogen or a more electropositive element or the removal of a more electronegative element or oxygen from a substance.

Below are some examples of reduction reactions:

- $2CH_2CH_2\ (g) + H_2\ (g) \rightarrow CH_3CH_3\ (g)$
- $2FeCl_3\ (aq) + H_2\ (g) \rightarrow 2FeCl_2\ (aq) + 2HCl\ (aq)$

Now, if we closely examine the above reaction, we would find that all the reactions above have both reduction and oxidation reactions.

The reaction in which $FeCl_3$ is getting reduced as electronegative element chlorine is being removed from it. At the same time, hydrogen is getting oxidized due to the addition of chlorine, an electronegative element, in the same reaction.

Oxidizing and Reducing Agents

- The substance (atom, ion and molecule) that gains electrons and is thereby reduced to a low valency state is called an oxidising agent.
- The substance that loses electrons and is thereby oxidised to a higher valency state is called a reducing agent.

Important Oxidizing Agents

- Molecules are made up of electronegative elements. Eg: O_2, O_3, and X_2 (halogens) • Compounds containing an element that is in the higher oxidized state. Eg: $KMnO_4$, $K_2Cv_2O_7$, HNO_3, $KClo_3$
- Oxides of metals and non-metals. Eg: MgO, CuO, CrO_3, P_4O_{10}
- Fluorine is the strongest oxidizing agent.

Important Reducing Agents

- All metals, for example, Na, Zn, Fe, Al
- A few non-metals, such as C, Hydrogen, S, P
- Hydracids, for example, HCl, HBr, HI, H_2S
- A few compounds contain an element in the lower oxidation state. Examples include, $FeCl_2$, $FeSo_4$, $SnCl_2$, Hg_2Cl_2
- Metallic hydrides, including NaH, LiH, CaH_2, etc.
- Organic compounds like HCOOH

Lithium is the strongest reducing agent in the solution, and caesium is the strongest reducing agent in the absence of water. The substances which act as oxidizing as well as reducing agents are H_2O_2, SO_2, H_2SO_3, HNO_2 and $NaNO_2$

Reduction Potential of a Half-reaction

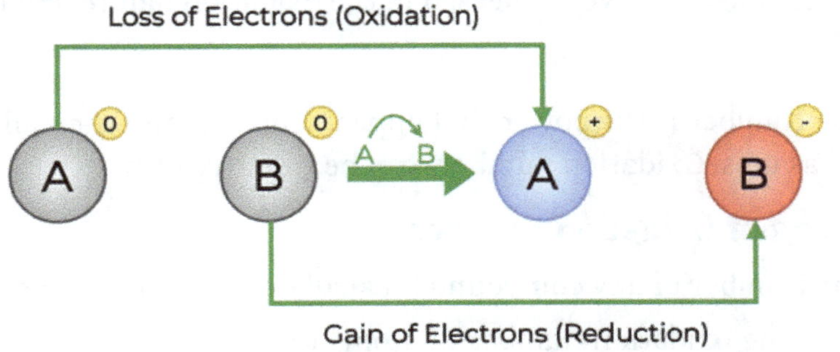

Fig. 7.5

Each half-reaction that makes up a redox reaction has a standard electrode potential. This potential equals the voltage produced by an electrochemical cell in which the cathode reaction is the half-reaction considered, whereas the anode is a standard hydrogen electrode.

This voltage produced by the half-reactions is called their reduction potentials (denoted by E^0_{red}). The value of the reduction potential of a half-reaction is positive for the oxidizing agents that are stronger than H+ and negative for the weaker ones.

Examples of the reduction potentials of some species are +2.866 V for F_2 and -0.763 V for Zn^{2+}.

Identification of Oxidizing and Reducing Agents

- If an element is in its higher possible oxidation state in a compound, it can function as an oxidising agent. Eg: $KMnO_4$, $K_2Cr_2O_7$, HNO_3, H_2SO_4 and $HClO_4$
- If an element is in its possible lower oxidation state in a compound, it can function as a reducing agent. Eg: H_2S, $H_2C_2O_4$, $FeSO_4$ and $SnCl_2$
- The compound will act as an oxidising agent if a highly electronegative element is in its highest oxidation state.
- The compound acts as a reducing agent if a highly electronegative element is in its lowest oxidation state.

Oxidation and Reduction Half Reactions

A reduction half and an oxidation half, which always occur together, make up redox reactions. While the oxidation half undergoes electron loss and an increase in oxidation number, the reduction half obtains electrons and the oxidation number decreases.

Mnemonic used are OIL stands for "oxidation is loss" and RIG stands for "reduction is gain," are simple ways to memorise this. The total number of electrons in a redox reaction remains unchanged. In the reduction half reaction, another species absorbs those that were released in the oxidation half reaction.

Oxidation Number

Oxidation Number is the power that appears on an atom when all other atoms are removed as ions. Oxidation numbers can be negative or positive.

Determination of Oxidation Number

Oxidation Number of any compound is calculated using the steps added below,

- Sum of the oxidation number of atoms in a neutral compound is equal to zero.
- Algebraic sum of the oxidation number of atoms in an ion is equal to the charge of that ion.
- Oxidation number of elements in their elementary state is equal to zero.
- Oxidation number of the element Fluorine is always equal to -1 in every compound.
- Oxidation number of Oxygen in every compound is -2, except in peroxides which are -1.

- Metal amalgam oxidation number is zero.
- Oxidation number of alkali metals is always +1 and the oxidation number of alkali earth metals is +2.
- For P-block elements, the highest oxidation number is always equal to their group number and their lowest oxidation number is equal to their group number minus 8.

For example: Lets find the Oxidation number of the compound K2Cr2O7 First consider the following oxidation number for the species present in the compound, K2 = 2(+1) {Since oxidation number of potassium is +1}

$$Cr2= 2(x)$$

O7 = 7(-2) {Since oxidation number of oxygen is -2}

By adding them,

$$2 + 2(x) - 14 = 0$$

$$2x - 12 = 0$$

$$2x = 12$$

$$x = 12/2$$

$$= +6$$

Therefore, oxidation number of chromium in K2Cr2O7 is +6

How to Balance Redox Reaction?

Chemical reaction are balanced according to "Law of Conservation of Mass". Chemical Equations which involve oxidation and reduction are balanced using following methods

- Oxidation Number Method
- Ion Electron Method (or Half Reaction Method)

Oxidation Number Method

Balancing chemical equations using the oxidation number method involves assigning oxidation numbers to each element in the reactants and products and then adjusting coefficients to ensure that the total increase and decrease in oxidation numbers are equal. The steps to balance a chemical equation using this method:

- **Step 1:** Write down the unbalanced chemical equation.
- **Step 2:** Assign oxidation numbers to each element in the reactants and products.

- **Step 3:** Identify which elements are undergoing oxidation (increasing oxidation number) and which are undergoing reduction (decreasing oxidation number).
- **Step 4:** Determine the change in oxidation number for each element undergoing oxidation or reduction.
- **Step 5:** Balance the equation by adding coefficients to ensure that the total increase and decrease in oxidation numbers are equal on both sides of the equation.

Example: Balance chemical equation by Oxidation Number Method

$$CuO + NH3 \rightarrow Cu + N2 + H2O$$

Ion Electron Method (or Half Reaction Method)

Ion-electron method, also known as the half-reaction method or the oxidation reduction (redox) method, involves balancing chemical equations by dividing the equation into two half-reactions: the oxidation half-reaction and the reduction half-reaction. Here's a step-by-step guide to balancing a chemical equation using this method:

- **Step 1:** Write down the unbalanced chemical equation.
- **Step 2:** Identify which elements are oxidized and which are reduced.
- **Step 3:** Write separate half-reactions for oxidation and reduction.
- **Step 4:** Balance each half-reaction individually by adding electrons (e-) to one side to balance the charge.
- **Step 5:** Multiply each half-reaction by appropriate coefficients to equalize the number of electrons transferred.
- **Step 6:** Add the balanced half-reactions together and cancel out any common terms.

Oxidation vs Reduction

- Oxidation occurs when a reactant loses electrons during a reaction. Examples of oxidation reaction is, C + O2 ® CO2
- Reduction occurs when a reactant acquires electrons during a reaction. Examples of reduction reaction is, N2 + 3H2 ® 2NH3

Uses of Oxidation and Reduction Reactions

Oxidation and Reduction Reactions are generally called redox reaction. Redox reactions are also used in a wide range of industries, including those in the biological, industrial, pharmaceutical, metallurgical, and agricultural fields. Examples of redox reactions in industrial processes include electroplating, the extraction of metals like aluminum, the production of chemical substances like caustic soda, the bleaching of wood pulp, metal corrosion, battery operation, and many more.

Redox reaction phenomena has been connected to a number of environmental challenges, including the ozone hole. So let's investigate the fundamental concept of redox reactions. Redox reactions are made up of two reactions that happen simultaneously: oxidation and reduction. that is, concurrently.

ELECTROCHEMICAL CELLS AND CELL POTENTIALS

Electrochemical cells permit this relative redox activity to be quantified by an easily measured property, potential. This property is more commonly called voltage when referenced in regard to electrical applications, and it is a measure of energy accompanying the transfer of charge. Potentials are measured in the volt unit, defined as one joule of energy per one coulomb of charge, V = J/C.

When measured for purposes of electrochemistry, a potential reflects the driving force for a specific type of charge transfer process, namely, the transfer of electrons between redox reactants.

The cell potential is a measure of the ability of the cell to do work and is related to the Gibbs free energy change (ÄG) of the electrochemical reaction.

A positive cell potential indicates that the reaction is spontaneous and can generate electrical energy, while a negative cell potential indicates that the reaction is non spontaneous and requires an external source of energy to occur.

The cell potential is an important parameter in many electrochemical processes, including batteries, fuel cells, and corrosion.

It can be measured experimentally using a voltmeter and can be predicted theoretically using the Nernst equation, which relates the cell potential to the concentrations of the reactants and products involved in the electrochemical reaction.

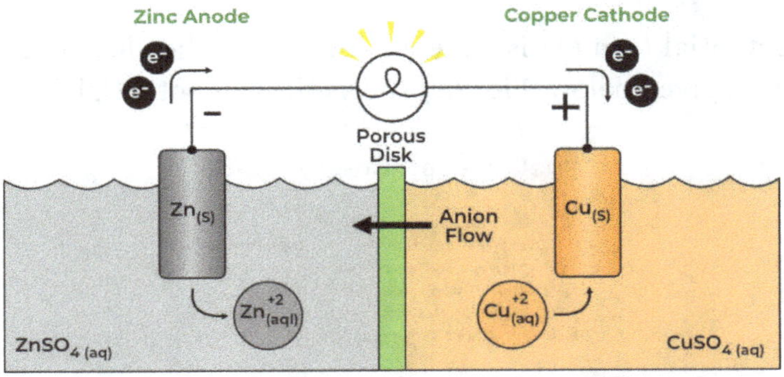

Figure 7.6 The structure of the Galvanic Cell

Figure 7.6 shows the structure of the Galvanic Cell for which we find the cell potential is calculated.

Types of Cell Potential

There are two main types of cell potential:

- Standard Cell Potential
- Non-Standard Cell Potential

Standard Cell Potential

- The cell potential that would be measured under typical conditions of 1 atm pressure, 25°C temperature, and 1 M concentration of all species participating in the electrochemical reaction is known as the standard cell potential (E°cell).
- Standard cell potentials are tabulated in reference books and used to forecast an electrochemical reaction's direction and determine the reaction's equilibrium constant. The Nernst equation can be used with standard concentrations to determine the standard cell potential.

Non-Standard Cell Potential

- Non-standard cell potential is the cell potential that is measured under non standard conditions, such as different concentrations of the species involved in the electrochemical reaction, temperature, and pressure.
- Nernst equation, which considers the impact of the non-standard conditions on the cell potential, relates the non-standard cell potential to the standard cell potential.
- Non-standard cell potentials can be used to predict the direction and extent of an electrochemical reaction under specific conditions and to calculate the reaction quotient of the reaction.

Cell Potential Formula

The cell potential formula is widely used to find the electric potential of the cell. The diagram given below (Fig. 7.7) shows the cell potential formula.

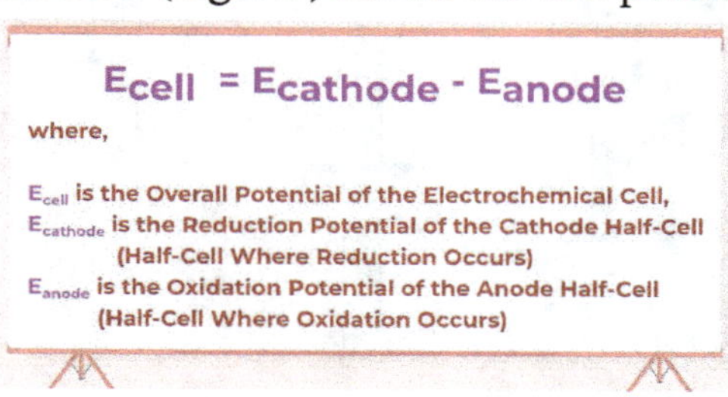

Fig. 7.7

To calculate the standard cell potential for a reaction

- Write the oxidation and reduction half-reactions for the cell.
- Look up the reduction potential, $E^{o}_{reduction}$, for the reduction half-reaction in a table of reduction potentials
- Look up the reduction potential for the reverse of the oxidation half-reaction and reverse the sign to obtain the oxidation potential. For the oxidation half-reaction, $E^{o}_{oxidation} = - E^{o}_{reduction}$.
- Add the potentials of the half-cells to get the overall standard cell potential. $E^{o}_{cell} = E^{o}_{reduction} + E^{o}_{oxidation}$

Example: Find the standard cell potential for an electrochemical cell with the following cell reaction.

$$Zn(s) + Cu^{2+}(aq) \rightarrow Zn^{2+}(aq) + Cu(s)$$

- Write the half-reactions for each process.

$$Zn(s)\ Zn^{2+}(aq) + 2\ e^{-}$$

$$Cu^{2+}(aq) + 2\ e^{-}\ Cu(s)$$

- Look up the standard potentials for the redcution half-reaction.

$$E^{o}_{reduction} \text{ of }_{Cu2+} = + 0.339\ V$$

- Look up the standard reduction potential for the reverse of the oxidation reaction and change the sign.

$$E^{oreduction} \text{ of }_{Zn2+} = -0.762\ V$$

$$E^{o}_{oxidation} \text{ of }_{Zn} = - (- 0.762\ V) = + 0.762\ V$$

- Add the cell potentials together to get the overall standard cell potential.

oxidation: $Zn(s)\ Zn^{2+}(aq) + 2\ e^{-}\ E^{o}_{ox.} = - E^{o}_{red.} = - (- 0.762\ V) = + 0.762\ V$

reduction: $Cu^{2+}(aq) + 2\ e^{-}\ Cu(s)\ E^{o}_{red.} = + 0.339\ V$

overall: $Zn(s) + Cu^{2+}(aq)\ Zn^{2+}(aq) + Cu(s)\ E^{o}_{cell} = + 1.101\ V$

NERNST EQUATION AND APPLICATIONS

The Nernst equation is very useful in analytical chemistry as well as in significant life processes like nerve conduction and membrane potential. Electrochemical cells and the Nernst equation calculates the pH of solutions, solubility products, constant equilibrium, and other thermodynamic properties, as well as potentiometric titrations and the calculation of cell membrane resting potentials.

The relationship between the potential of the electrode and the potential of the reference electrode is thus provided by this equation. Additionally, it also helps to compute the Gibbs free energy and forecast the spontaneity of an electrochemical reaction.

The electrical potential disparity across the cell membrane of all living cells is called the membrane potential, the inner part of the cell being negative compared to the outside. The magnitude of the membrane potential varies from cell to cell and in an exceptional cell following its functional state. For example, a nerve cell has a membrane potential of -70mv at rest, but the membrane potential drops to about +30mv when excited. The membrane potential at rest is called the resting potential. RMP is basically due to-

- Uneven distribution of ions across the cell membrane due to its selective permeability.
- Due to the combined effect of forces acting onions. This origin of RMP is dependent

Selective Permeability of Cell Membrane

The cell membrane is selectively permeable, i.e. it is freely permeable to K+ and Cl– , from the medium to Na+, and is impermeable to proteins and organic phosphates which are negatively charged ions. Usually, the intracellular cation is K+ and major intracellular anions are proteins and organic phosphate. General extracellular cation is Na+ and anion is Cl–.

The presence of gated protein channels in the cell membrane is responsible for the variable permeability of ions. The forces executive on the ions across the cell membrane Production of variations in the membrane potential. The magnitude of forces acting thenceforward the cell membrane on each ion can be analyzed by the Nernst equation.

Concentration gradient: The Donnan effect results in an uneven distribution of diffuse ions across the cell membrane, in the form of additional diffuse cations, resulting in a concentration gradient.

Electrical gradient: As a consequence of concentration gradient cation K+, will try to disseminate back into ECF from ICF. But it is counteracted by an electrical gradient which will be created due to the impendence of nondiffusible anions within the cell. The membrane potential at which the electric force is equal in magnitude but opposite in the direction of the concentration force is called the equilibrium potential for that ion. The magnitude of the equilibrium potential is determined by the Nernst equation.

$$E_m = -RT/ZF \text{ In } C_{in}/C_{out}$$

37 °C at normal body temperature, substituting for the constants (R, T, and F) and converting to the normal logarithm,

$$E_m = -61.5 \log C_{in}/C_{out}$$

Nernst equation

The Standard electrode potentials are measured at their standard states when the concentration of the electrolyte solution is fixed as 1M and the temperature is 298 K. Despite this, in actual practice, electrochemical cells do not always have a fixed concentration on the electrolyte solution. The electrode potential depends on the concentration of the electrolyte solution. The Nernst equation gave a relationship between the electrode potential and the concentration of electrolyte solutions, known as the Nernst equation. For a general electrode:

$$M^{n+} + ne^- \rightarrow M$$

$$E_{cell} = E_{cell} - (RT/nF)\ln Q$$

$$E_{cell} = E^\circ_{cell} - (RT/nF)\ln([C]^c[D]^d / [A]^a[B]^b)$$

where,

- R = The gas constant ($8.314 JK^{-1}mol^{-1}$)
- F = Faraday constant ($96{,}500 Cmol^{-1}$)
- T = Temperature in kelvin
- Q = Reaction quotient
- n = Total number of moles of electrons translocate

It should be remembered that when writing the Nernst equation for the overall cell reaction, the log term is the same as the expression for the equilibrium constant for the reaction. The relation of both is similar.

$$E_{cell} = E^\circ_{cell} - (2.303\ RT/nf)\ \ln([C]^c[D]^d / [A]^a[B]^b)$$

$$E_{cell} = E^\circ_{cell} + 2.303\ RT\ [A]\ [B]\ln([A]^a[B]^b / [C]^c[D]^d)$$

Similarly, for the electrode reaction:

$$M^{n+} + ne^- \rightarrow M$$

The Nernst equation is

$$E_{cell} = E_{cell} - (2.303\ RT/nf)\log[1/M^{n+}]$$

$$\text{or } E_{cell\,=}\ E_{cell\,+}\ (2.303RT / nF)\log[1/M^{n+}]$$

when T = 273K, F=96500 Cmol-1, R=8.314JK-1mol-1 and concentration of solid M is taken as unit

$$E_{cell} = E_{cell} - (0.059/n)\log[1/M^{n+}]$$

Redox Reactions and Electrochemistry 117

Relationship between equilibrium constant and standard potential of a cell

$E_{cell} = E_{cell} - (2.303\ RT/nF)\log[K_c]$, [$K_c$ = equilibrium constant]

At equilibrium, Ecell =0

$$E_{cell} = (2.303\ RT/nF)\log[K_c]$$

$$K_c = \text{antilog}[nE_{cell}/0.0591]$$

Limitations of Nernst Equation

- The Nernst equation applies exclusively because no current flows through the electrodes. When current flows, the movement of ions at the electrode surface changes, and the conditions of excess potential and resistance loss contribute to the measured potential.
- At very low concentrations of commutation potential-determining ions, the potential approach ± is found using the Nernst equation. This is corporeally useless, because, underneath such a situation, the commutation current density is reduced and tends to control the electrochemical behaviour of the system more than other effects.
- Since the active coefficients are close to unity in dilute solutions, the Nernst equation can be expressed in the directly implicit form of the concentration. But in the case of higher concentrations, the actual activities of the ions must be used. This creates complexity for the use of the Nernst equation because estimating the non-ideal activities of ions usually requires experimental measurements.

Application of Nernst Equations

Using the equation, one can determine:

- Under any circumstances, the single electrode reduction or oxidation potential • Standard electrode potentials
- Comparing the relative ability as a reductive or oxidative agent. • determining whether it is feasible to combine such single electrodes to create electric potential.
- Emf of an electrochemical cell
- unknown concentrations of ions
- The Nernst equation can also calculate the pH of solutions and the solubility of salts that are only marginally soluble.
- Used in the marine environment
- Helpful in figuring the potential of an ion with the charge "z" over the membrane.

CORROSION AND PROTECTION OF METALS

Corrosion is described as the chemical and electrochemical destruction of metals or alloys by the surrounding damp and humid environment. Corrosion is an oxidation process since it occurs in the presence of oxygen. Almost all metals corrode, however, the rate of corrosion varies depending on the metal.

Metals generate stable compounds such as oxides, hydroxides, and sulphides as a result of corrosion, which is a spontaneous and irreversible process.

Corrosion Definition

The degradation of pure material when they react with moisture, air or acid is called Corrosion. Corrosion is a redox reaction that causes the loss of pure metal by converting them into oxide and thus decreasing its strength and other physical properties.

Factors Influencing Corrosion

Various factors influencing the corrosion of any material are:

- **Position of Metals in Electrochemical Series:** In corrosion, the electrochemical series is extremely important. Reactive metals lose electrons more quickly and corrode more quickly. Iron, for example, is quickly oxidised, but gold is not because gold is less reactive than iron.
- **Impurities in Metals:** In general, the presence of contaminants in metals accelerates corrosion because these impurities operate as the microscopic electrochemical cell that causes corrosion.
- **Presence of Electrolytes:** Because they carry ions, dissolved salt in water acts as an electrolyte. Corrosion is accelerated when electrolytes are present in water.
- **Concentration of Oxygen:** Corrosion is accelerated by an increase in oxygen content. An anode is a location with less oxygen concentration, whereas a cathode is a region with a higher oxygen concentration. Corrosion happens as a result.
- Humidity in Weather: When the weather is humid, the presence of moisture with a high temperature accelerates corrosion because ions gain energy and begin moving quicker in a higher temperature, causing them to collide more frequently.
- Temperature: The rate of Corrosion increases with temperature.

Rate of Corrosion

The formation of an Oxide layer on metal is explained by the Deal Grove model. With the help of this model, we can predict and control the formation of Oxide layers even in adverse situations. Since corrosion lead to the loss of metal, one can ask what is lost. The answer is the mass of the metal is lost. Hence, to measure how much corrosion occurred one should measure the mass lost from the metal. To measure the mass lost, weigh a clean pure piece of metal and then expose it to the environment for a definite period. After corrosion had occurred clean the corroded surface and again measure the mass of the metal block.

We can easily calculate the rate of corrosion (R) using the following formula,

$$R = kW/rAt$$

where,

- k is the proportionality constant
- W is the weight loss of the metal in time t

Redox Reactions and Electrochemistry 119

- A is the surface area of the metal exposed
- r is the density of the metal (in g/cm^3)

Corrosion Types

There are various types of corrosion and all of them are explained below,

Uniform Corrosion

Corrosion of this sort is fairly prevalent. This kind of corrosion results in a homogeneous loss of metal throughout a whole surface. Rusting, tarnishing of silver, and so forth are examples.

Pitting Corrosion

Pitting corrosion is a type of localised corrosion that damages the metal surface, resulting in patches or pits. In neutral or acid solutions containing halides, such as saltwater containing NaCl, pitting corrosion can be observed in stainless steels.

Crevice Corrosion

Crevice corrosion is caused by a stagnant solution in crevices around the edges of metal surfaces, such as nuts and rivet heads. It can happen when two metals collide or when a metal collides with a non-metal. Corrosion on metal or non-metal gaskets, couplings, and joints, for example.

Galvanic Corrosion

When two dissimilar materials are connected and brought into electrical contact with a corrosive electrolyte, galvanic corrosion occurs. In galvanised iron, for example, a sheet of iron or steel is coated with zinc.

Stress Corrosion Cracking

It's a form of corrosion in which fractures appear on the metal's surface as a result of the combined effects of tensile strength and the corrosive environment. Stress Corrosion Cracking (SCS) is a kind of corrosion that occurs in aviation components.

Intergranular Corrosion

It's a sort of corrosion that happens around the grain boundaries of a metal or alloy that contains impurities, rather than inside the metal or alloy. Die-cast zinc alloys containing aluminium exposed to steam in a maritime environment display intergranular corrosion.

Dealloying

It's a rare sort of corrosion that only affects particular metals, such as copper alloys and grey cast iron. The alloy metal loses its reactive element and creates a more stable corrosion-resistant element in a porous condition during the dealloying process. De nickelification, de-zincification, and graphitic corrosion are all instances of dealloying.

Erosion Corrosion

Erosion Corrosion is a mechanically induced breakdown of a metal surface. In a corrosive environment, it is caused by the relative mobility of fluid particles and solid surfaces. Burrs left at the cut tube ends of pipelines, for example, might disrupt smooth water flow, causing localised turbulence and high flow velocities, resulting in erosion corrosion.

Various Examples of Corrosion

Metals with a higher reactivity series, such as iron, copper, and zinc, corrode more quickly. Corrosion can manifest itself in a variety of ways.

Rusting of Iron

Rusting is the process related to iron metal, the corrosion of iron in the presence of air and water is called Rusting. It results in the formation of a Reddish-Brown surface. When iron oxidises is formed in the presence of oxygen and water we say that rusting has occurred. The chemical formula of the rust is Fe2O3.xH2O

Though rusting of iron occurs naturally, we can represent the formation of rust as an electrochemical reaction as,

At Cathode

$$H^+ + e^- \rightarrow H$$

$$4H + O_2 \rightarrow 2H_2O$$

$$4H^+ + O_2 + 4e^- \rightarrow 2H_2O$$

At Anode:

$$2Fe \rightarrow 2Fe^{2+} + 4e^-$$

Oxidation Half-reaction

$$Fe(s) \rightarrow 1/2Fe_2^{+}(aq) + 2e^-$$

Reduction Half-reaction

$$4H^+ + O_2 + 4e^- \rightarrow 2H_2O$$

Overall Cell reaction

$$2Fe(s) + 4H^+ + O2 \rightarrow 2Fe^{2+}(aq) + 2H_2O$$

$$2Fe_3+ + 3O_2 \rightarrow 2Fe_2O_3$$

$$Fe_2O_3 + xH_2O \rightarrow Fe_2O_3.xH2O(rust)$$

Corrosion of Copper

When exposed to moist air, copper metal or alloys of copper corrode, i.e. they form a pale-looking copper oxidises layer, and the shiny copper surface tarnishes, i.e. it turns dark brown or black, then green.

The blue-green colour of the "Statue of Liberty" is an example of corrosion of copper metal. It was not painted the same colour as it is now when it was built. Rather, it was brown in colour. The copper oxidised over time and reacted with sulphur trioxide, carbon dioxide, and water in the atmosphere to generate a new chemical that gave the Statue of Liberty its distinctive blue-green patina.

The following is the chemical equation involving the corrosion of metal is,

$$Cu + H_2O + CO_2 + O_2 \rightarrow Cu(OH)_2 + CuCO_3$$

Tarnishing of Silver

Tarnishing is the process of the silver metal becoming black due to a covering of silver sulphide. Chemical interaction between the silver and sulphur-containing compounds in the air causes the silver to tarnish. Silver sulphide is formed when silver is mixed with sulphur.

The following is the chemical equation showing the tarnishing of silver,

$$2Ag + H2S \rightarrow Ag_2S + H_2$$

Prevention of Corrosion

Corrosion is an unwelcome phenomenon that must be avoided since it causes significant damage to metallic constructions such as bridges, sculptures, monuments, metal utensils, and so on.

All corrosion prevention methods are based on the same basic concept: cutting off the metals' water and air supply. The following are some of the corrosion prevention measures:

- **Painting:** Painting iron items, such as gates and rails, prevents metal from being exposed to air and water. Painting metallic items protect them from corrosion.
- **Oiling and Greasing:** Oiling and greasing, like painting, provide a protective coating on the metal surface that protects it from corrosion.
- **Galvanisation:** Galvanisation is the process of coating iron and steel things with a thin layer of zinc to protect them. The galvanised products are protected against corrosion or rusting even after the zinc coating has been broken.
- **Alloying:** Alloys are made up of a homogeneous combination of metals and non metals. We may achieve the required characteristics of metals by alloying them. Iron, for example, is a very useful metal, yet it rusts fast. As a result, when this iron is combined with Nickel and Chromium, stainless steel is created. Stainless steel is now widely utilised in the manufacture of kitchenware.

8

ORGANIC REACTION MECHANISMS

Organic chemistry is the chemistry of carbon compounds except for oxides of carbon and metal carbonates. Carbon has the uncommon characteristic of forming strong bonds with many other elements, particularly with other carbon atoms, to form chains and rings, giving rise to millions of organic molecules. Carbon compounds are essential for the survival of life on Earth because of this distinguishing property. Proteins, DNA (deoxyribonucleic acid), and other complex organic substances provide chemical, structural, or genetic functions in living beings.

A chemical equation only shows the starting and final products of a reaction; it rarely shows how the process proceeds. Some reactions occur through intermediates, which may or may not be separated depending on their stability.

Mechanism is the complete step-by-step description of the order in which bonds break and bonds form, to give the observed products.

ELECTRON MOVEMENT IN ORGANIC REACTIONS

Curved arrows are used to depict electron transport in chemical processes. The curving arrow is a handy indication for showing bonding changes caused by electronic redistribution during the reaction. A covalent connection is created when two atoms share a pair of electrons. The many methods in which the electron-pair travels are depicted below using curved arrows. The arrow begins at the point when the electron-pair is displaced and ends at the point where the electron-pair travels.

- Shifting of electron-pair from p-bond to adjacent bond position.

- Shifting of electron-pair from p-bond to adjacent atom.

- Shifting of electron-pair from atom to adjacent bond position.

For Example: In the case of ethane when reacting in presence of UV light, due to the electron shifting, two methane molecules are obtained as shown below,

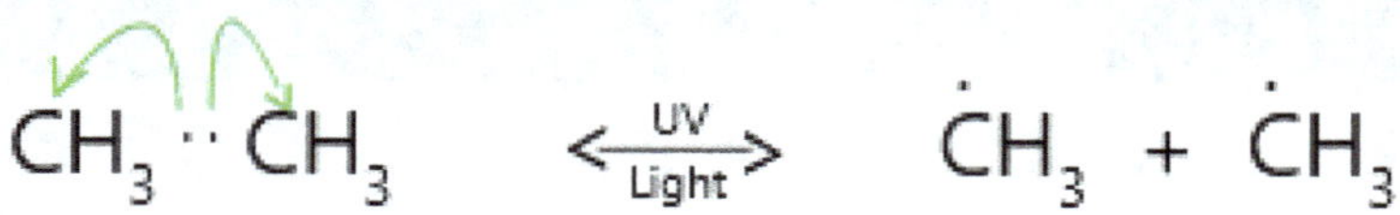

Bond Fission

In any chemical reaction, when a reactant is converted into products one or more bonds in the reactant are broken and new bonds are formed. The process of breaking or cleavage of a covalent bond is known as bond fission.

Now, the bond fission takes place in two ways as mentioned below,

Homolytic fission

The symmetrical breaking of a covalent bond between two atoms such that each atom acquires one electron of the shared pair is called homolytic fission or homolysis.

Such fission takes place in the presence of ultraviolet light or at high temperatures. The cleavage of a bond results in the formation of free radicals. A free radical is a neutral species (atom or group) that contains an unpaired electron. Homolytic fission can be shown as,

The movement of a single electron is shown by a half-headed curved arrow (Fishhook). Free radicals have transitory existence i.e. they are short-lived and are highly reactive. They are paramagnetic.

Generally, a covalent bond between two atoms of the same element or two atoms having nearly the same electronegativity breaks in this manner. For example, Organic reactions which proceed by homolysis are called free radical or homopolar or non-polar reactions as these reactions take place in a non-polar solvent. Homolysis generally occurs in the gaseous phase in presence of sunlight or ultra-violet light or in the presence of catalysts such as hydrogen peroxide.

Heterolytic fission

The unsymmetrical breaking of a covalent bond between two atoms such that the more electronegative atom acquires both the electrons of the shared pair is called heterolytic fission or heterolysis.

Such fission takes place in the presence of a polar solvent. The cleavage of a bond results in the formation of ions. One of the ions has a sextet electronic structure and a positive charge called a cation and the other ion has a valence octet with at least one lone pair and a negative charge called an anion.

Heterolytic fission can be shown as,

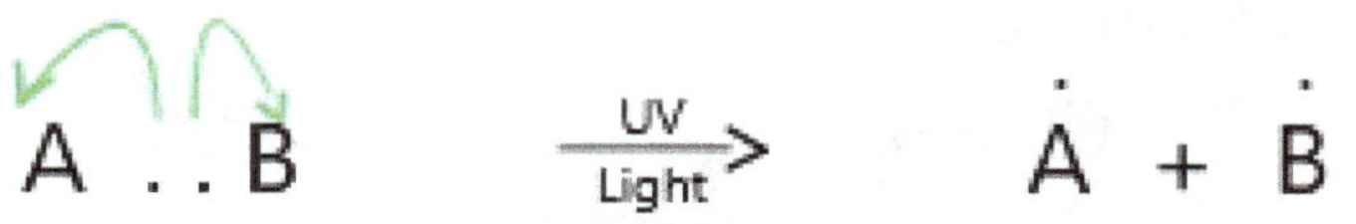

⦿ where B is more electronegative than A.

If A is more electronegative than B, then the fission will be shown as,

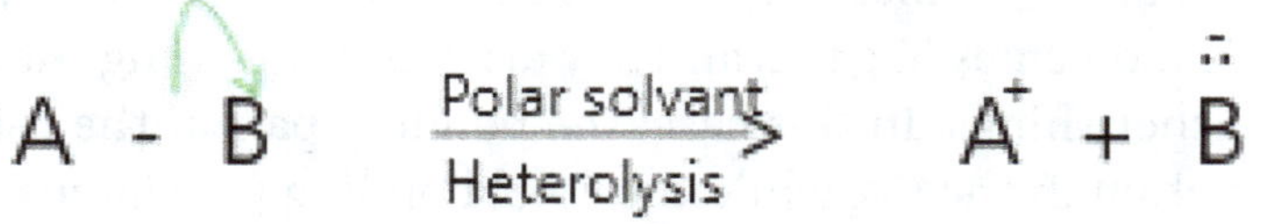

The ions formed are unstable and reactive. An example of heterolysis is,

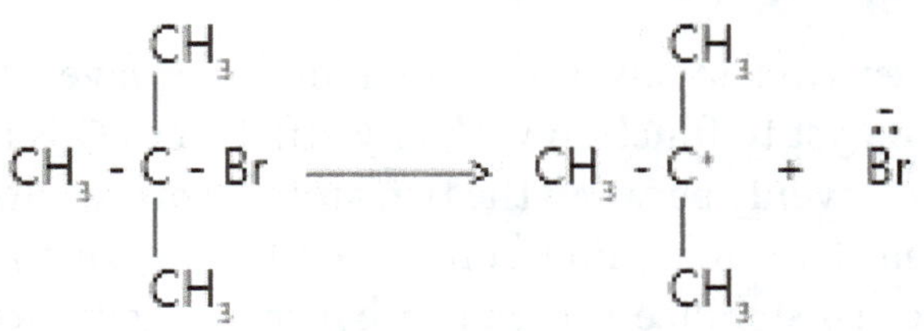

The species in which a carbon atom possesses a sextet of electrons and a positive charge is called carbocation or carbonium ion. A carbocation is electron deficient. In the C-Br bond, the bromine atom is more electronegative than the carbon atom and hence the electron pair is retained by the bromine atom on fission.

But consider the reaction:

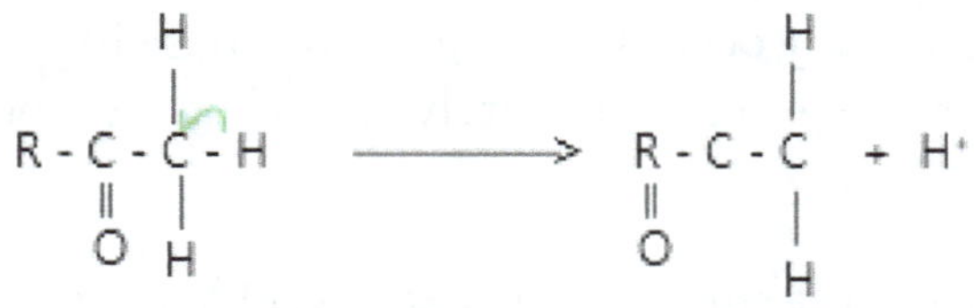

In the C-H bond, the carbon atom is more electronegative than the hydrogen atom and hence the electron pair is retained by the carbon atom on fission. The species in which a carbon atom possesses an octet of electrons and a negative charge is called carbanion. Carbanions are reactive as they are unstable. Organic reactions which proceed by heterolytic are called ionic or heteropolar or simply polar reactions as these reactions take place in a polar solvent. Heterolysis is uncommon in a gaseous state.

Chlorination of Methane

In the chlorination of methane, we see a particular property of hydrogen atoms in organic molecules in general. They are labile, meaning they can be easily exchanged for other things. In this case, the bonding pair of the C-H bond moves to one Chlorine atom in the Cl2 while the Cl-Cl bonding pair jumps to form a new bond between the other Cl and the Carbon atom.

Cracking Reaction

The cracking reaction involves the homolitic cleavage of a C-C bond. This reaction needs a catalyst to function with any efficiency. This means that the two electrons are divided evenly between the two atoms. This results in two fragments, each with a lone electron (dangling bond). The two fragments are radicals and are not very stable. To stabilize the pair, a hydrogen homolitically cleaves from a carbon neighboring one of the two carbons used to be bonded together leaves that carbon and migrates over to the other fragment to complete the bonding on that fragment. The two carbons on the fragment from which the hydrogen left now form a double C=C bond and all electrons are paired up again. See the diagram below for a visual representation.

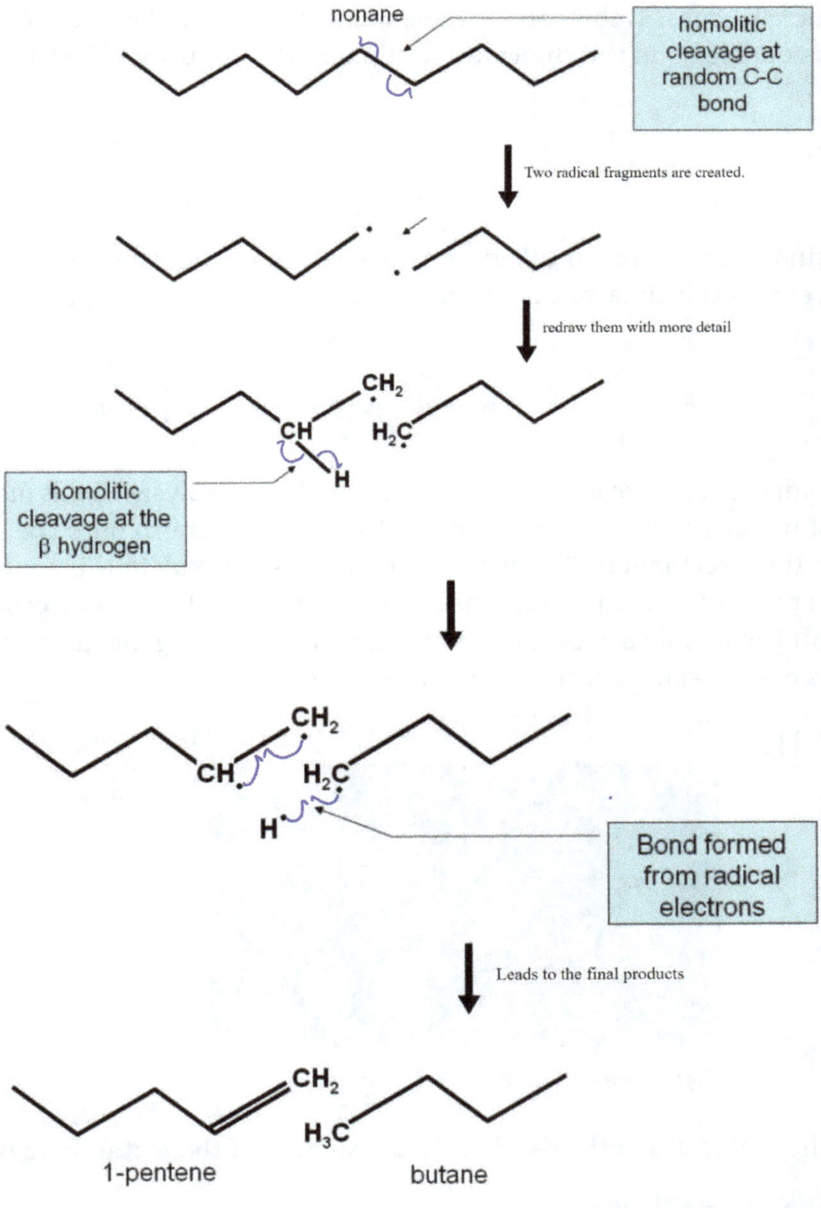

Note that the double bond could have formed on the 4-carbon fragment as well, to produce 1-butene and pentane in more or less equal amounts. So, each bond broken in a cracking reaction could form two sets of compounds for a total of 4 products.

When HCl is added to an alkene or an alkyne, the pi electrons from the multiple bond form a new bond with the proton on the HCl, leaving behind a carbocation where the ene was and a chloride ion, which then combine in a second step to make the product. Because, in this case, the ene was symmetric, it does not matter which side of the double bond the chloride ends up on, the product will be the same.

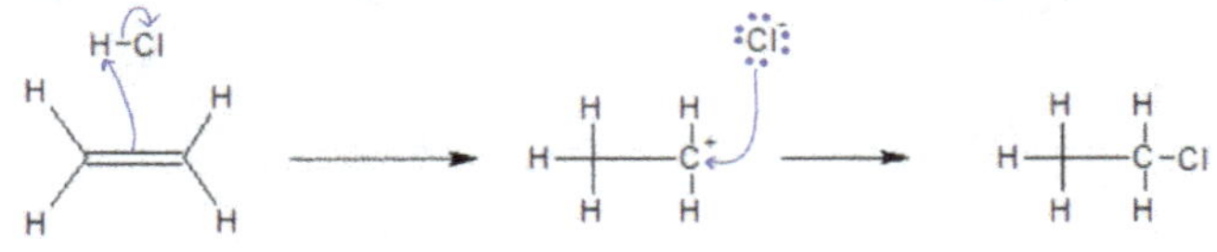

Bromine addition to an alkene or alkyne works the same way as the HCl addition, except this time, one Br atom ends up on each carbon from the ene group.

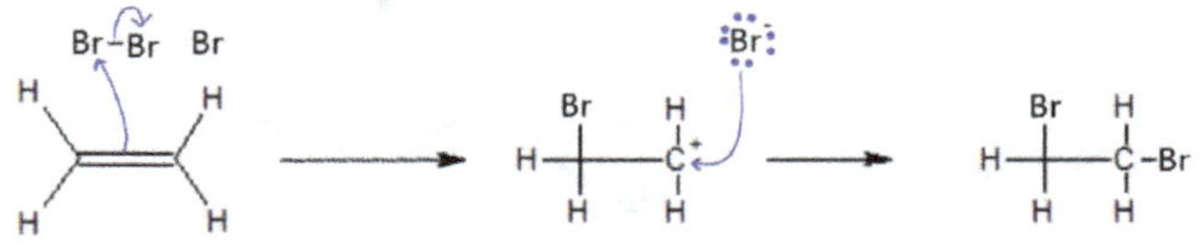

The hydrogenation reaction will not occur without catalysis. In this mechanism, a sheet of metal acts as a catalyst. The molecules both adsorb onto the surface of the metal (blue rectangle). The hydrogen bonds easily break in this condition and they then migrate to the pi electrons in the alkene (or alkyne). The product then desorbs off the metal surface. This is an example of heterogeneous catalysis. The catalyst is not the same phase as the reactants (Fig. 8.1).

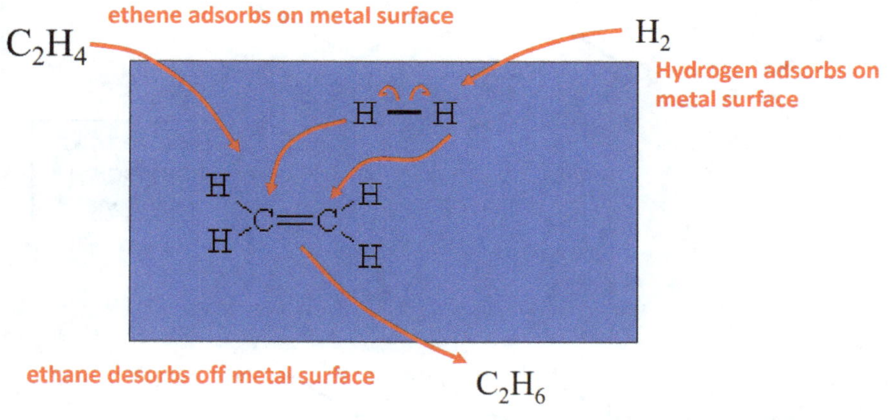

Fig. 8.1 The molecules both adsorb onto the surface of the metal (blue rectangle)

Markovnikov Reactions

In the addition of an acid (say, HCl) to an 'ene' group, the proton (electrophile) first attacks the p electrons of the double bond, forming a new s bond with one of the carbons, leaving the other carbon with one too few electrons (carbocation). Then, in the second step the nucleophile Cl– attacks the carbocation, filling it's empty orbital and forming a new bond.

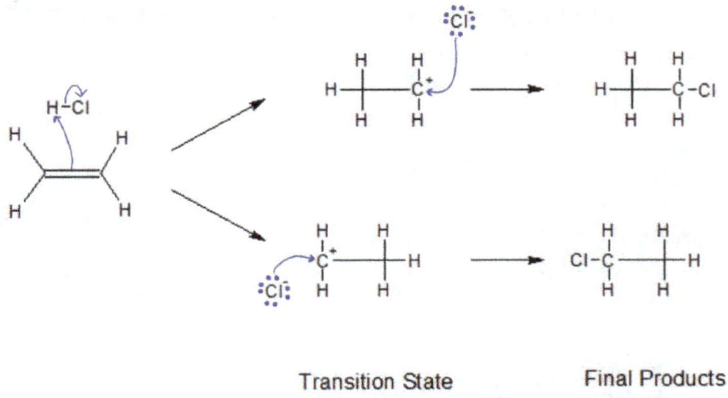

Fig. 8.2 Markovnikov addition of HCl to a symmetric alkene. Results in equal amounts of two identical products, so one product.

The transition state in both these first steps have to be stable enough to stick around long enough for the Nucleophile to attack. If it were to decompose too quickly, the reaction would never reach completion. Since both sides of the 'ene' group are identical in this example, both products will be equally likely. Here, they are indistinguishable anyway.

In the second step, the reactive part of the transition state is the carbocation (it has the charge). Anything we do to stabilize the carbocation (reduce the charge) will result in that particular reaction path being favored because it will help the transition state 'stick around' long enough for the second step to happen.

If we put on one of the carbons a substituent that pushes electron density onto that carbon, then that carbon will be stabilized if it becomes the carbocation. If, we put an electron-withdrawing (highly electronegative) substituent on a carbon then it will be destabilized as a carbocation. It turns out that R groups (alkyl groups) are electron donating (they have lower electronegativity than a Carbon with double bonds) while a halogen would be electron withdrawing (it has high electronegativity).

Thus, for a hydrocarbon-only compound with an asymmetrically substituted 'ene' group, the carbon with more R groups will be stabilized relative to the carbon with more hydrogens (since the R group will push electron density onto the carbocation, distributing the positive charge and stabilizing it.)

In the case of non-alkyl groups on the 'ene' carbon, we must first see if it is electron donating (like an alkyl) or electron withdrawing. If we have Cl as a substituent, the opposite reaction path becomes favored. The highly electronegative Cl tries to pull electron density away from the already positively charged carbon. The carbocation will be thusly destabilized, i.e., the transition state will quickly dissociate back to the starting materials, making the second step of that reaction path less likely to occur.

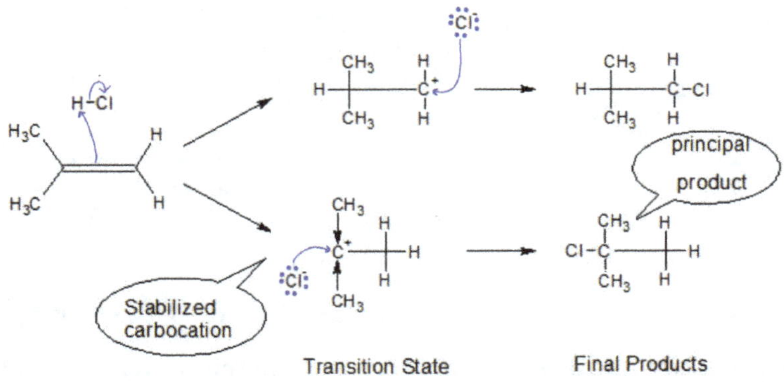

Fig. 8.3 Markovnikov addition of HCl to a methyl-substituted alkene. The methyl side is stabilized in the transition state so "the hydrogen goes to the side with the most hydrogens". Markovnikov's rule.

Markovnikov addition of HCl to a chloro-substituted alkene. The carbocation is destabilized by the electron withdrawing chloro group so the opposite product dominates. (Anti Markovnikov rule).

SUBSTITUTION AND ELIMINATION REACTIONS

A nucleophilic substitution reaction is not the only possible reaction for alkyl halides and other substrates with a good leaving group. These substrates can also undergo elimination reactions. In an elimination reaction (Fig. 8.4), a small molecule (XY) is removed from two adjacent atoms of the reactant, and a multiple bond is formed in the product:

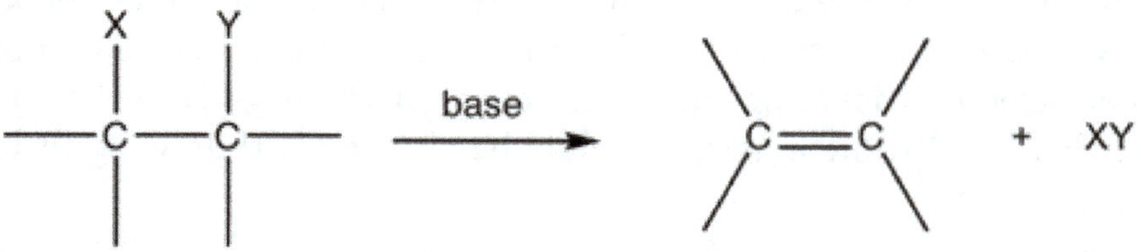

Fig. 8.4 General equation of elimination reaction

For an elimination reaction of alkyl halides, the major product alkene is produced together with the small HX (X is halogen) molecule, which is the side inorganic product. Such reaction with removal of a proton and a halide ion is called dehydrohalogenation (Fig. 8.5). Dehydrohalogenation is a commonly applied method for the synthesis of alkene.

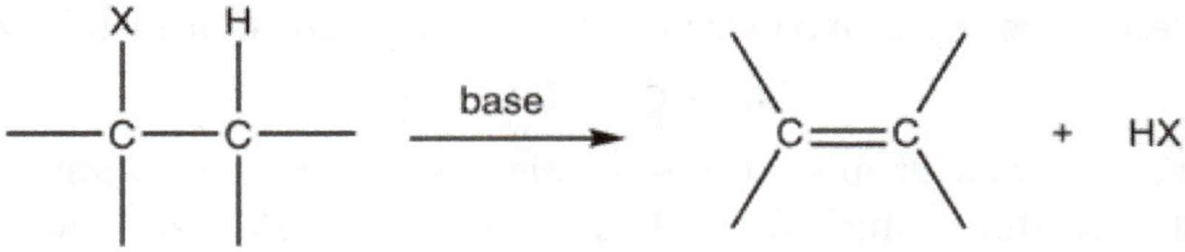

Fig. 8.5 Dehydrohalogenation of alkyl halide

In the discussions of elimination reactions, the carbon atom that bonded to the leaving group (halogen for alkyl haide) is called the alpha (a) carbon atom, and the carbon atom adjacent to a-carbon is called the beta (b) carbon atom (Fig. 8.6). In dehydrohalogenation, the hydrogen atom on â-carbon is eliminated together with halogen, as HX; therefore, the reaction is often called b-elimination or 1,2-elimination. (The number 1,2- indicates that the atoms being removed are on two adjacent carbons and not necessarily C1 and C2 atoms.)

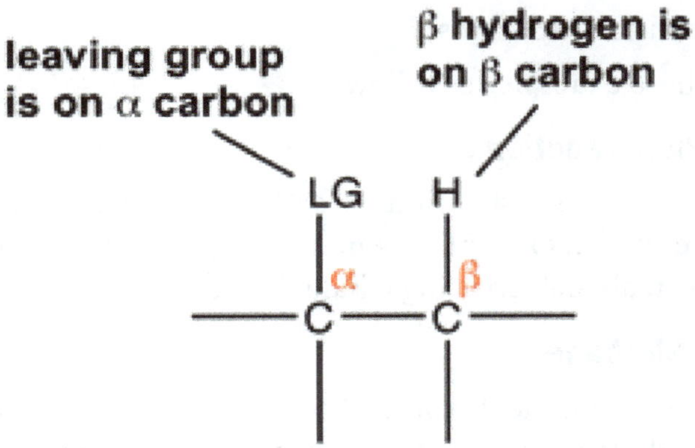

Fig. 8.6 Alpha and Beta carbon

Beside the substrate, a base is required for elimination reactions other than the nucleophile for substitution. Similar to substitution reactions, elimination reactions also have different rate laws, and therefore involve different mechanisms. The elimination mechanisms are E1 and E2. You may expect that E1 is a unimolecular reaction with the first order rate law and E2 is the second order bimolecular reaction. That is correct. We will go through the mechanism in detail first, then compare between elimination and substitution.

Substitution Reaction

Substitution reactions are fundamental chemical reactions in chemistry, and they involve the exchange of one atom or group of atoms (functional group) with another in a molecule.

Considering the reaction between methane (CH_4) and chlorine (Cl_2),

$$CH^4 + Cl_2 \rightarrow CH_3Cl + HCl$$

This is an example of substitution reaction as one hydrogen atom in methane with a chlorine atom, which forms hydrochloric acid (HCl) and chloromethane (CH3Cl).

Substitution Reaction Definition

A substitution reaction is a chemical reaction where one functional group or atom within a molecule is replaced by another functional group or atom.

Examples of Substitution Reactions

Some of the most common examples of substitution reactions are:

- Alkane Substitution Reactions
- Acyl Substitution Reactions

Let's discuss these examples as follows:

Alkane Substitution Reactions

Alkanes experience substitution reactions because they are saturated hydrocarbons. One typical example is when UV light causes a halogen to replace a hydrogen atom in an alkane, creating a haloalkane.

Halogenation of Methane

Halogenation of methane involves the substitution of hydrogen atoms in methane (CH_4) with halogen atoms, typically chlorine (Cl_2) or bromine (Br_2).

$$CH_4 + Cl_2 \rightarrow CH_3Cl + HCl$$

Bromination of Ethane

Bromination of ethane involves the substitution of hydrogen atoms in ethane (C_2H_6) with bromine (Br_2) atoms. This reaction is a type of substitution reaction and typically requires the presence of heat or ultraviolet (UV) light to initiate it.

$$C_2H_6 + Br_2 \rightarrow C_2H_5Br + HBr$$

Chlorination of Propane

The chlorination of propane involves the substitution of hydrogen atoms in propane (C_3H_8) with chlorine (Cl_2) atoms. This reaction is a type of substitution reaction and usually requires the presence of heat or ultraviolet (UV) light to initiate it.

$$C_3H_8 + Cl_2 \rightarrow C_3H_7Cl + HCl$$

Acyl Substitution Reaction

An acyl substitution reaction is the process by which an acyl group is replaced in a compound. These are common reactions in organic synthesis, especially when amides and esters are formed.

Nucleophilic Acyl Substitution in Carboxylic Acids

Nucleophilic acyl substitution is a type of organic reaction that involves the substitution of an acyl group (RCO–) in a carboxylic acid or a derivative of a carboxylic acid with a nucleophile.

General reaction scheme for nucleophilic acyl substitution in carboxylic acids is given as:

$$R\text{-}COOH + Nu^- \rightarrow R\text{-}CO\text{-}Nu + OH^-$$

Let's consider an example for the same.

$$RCOOH + NH_3 \rightarrow RCOONH_2 + H_2$$

In this reaction, ammonia (NH_3) replaces the hydroxyl group (-OH) in a carboxylic acid (RCOOH) to form an amide ($RCOONH_2$).

Nucleophilic Acyl Substitution in Acid Chlorides

Nucleophilic acyl substitution involves the substitution of the acyl group (RCO-) in an acid chloride (RCOCl) with a nucleophile.

General reaction for this is given as:

$$RCOCl + Nu^- \rightarrow RCONu + Cl^-$$

Let's consider and example for the same.

$$RCOCl + HOCH_3 \rightarrow RCOOCH_3 + HCl$$

Methanol (HOCH3) reacts with an acid chloride (RCOCl) to produce an ester (RCOOCH3) and hydrogen chloride (HCl).

Nucleophilic Acyl Substitution in Esters

Nucleophilic acyl substitution involves the substitution of the acyl group (RCO-) in an ester (RCOOR') with a nucleophile.

General reaction for the same is given as:

$$RCOOR' + Nu- ® RCONu + R'O-$$

Let's consider an example for the same.

$$RCOOCH_3 + NH_2NH_2 \rightarrow RCOONHNH_2 + CH_3OH$$

Hydrazine (NH2 NH2) replaces the methoxy group (-OCH_3) in an ester ($RCOOCH_3$) to form a hydrazide ($RCOONHNH_2$) and methanol (CH_3OH).

Classification of Substitution Reactions

Substitution reactions can commonly be classified as follows:

- Nucleophilic Substitution
- Electrophilic Substitution

Other than this there are some more classification of substitution reaction given as: • Free Radical Substitution

- Organometallic Substitution

Nucleophilic Substitution Reactions

In a nucleophilic substitution reaction, a nucleophile (electron-rich species) replaces a leaving group (a group that can depart as a stable ion or molecule) from a molecule. The nucleophile attacks the electrophilic center of the molecule, leading to the substitution of the leaving group with the nucleophile.

Common examples of nucleophilic substitution reactions include SN1 (substitution nucleophilic unimolecular) and SN2 (substitution nucleophilic bimolecular) reactions.

Nucleophilic Substitution Examples

Some of the common example all types of substitution reaction are listed as follows:

- **Examples of SN_1 Reaction**

Tertiary Butyl Chloride ® Tertiary Butanol (tert-butanol) + Chloride ion (Cl-)

- **Examples of SN_2 Reaction**

Chloromethane (CH3Cl) + Hydroxide ion (OH-)® Methanol (CH3OH) + Chloride ion (Cl-)

Some Other Examples

Some other examples include halogenation of alkane i.e.,

- Methane (CH_4) + Chlorine Gas (Cl_2) → Chloromethane (CH_3Cl) + Hydrochloric Acid (HCl)
- Ethyl Bromide (CH_3CH_2Br) + Hydroxide ion (OH^-)→ Ethanol (CH_3CH_2OH) + Bromide ion (Br–)

Electrophilic Substitution Reactions

In an electrophilic substitution reaction, an electrophile (electron-poor species) replaces an atom or group in a molecule. This type of reaction commonly occurs in aromatic compounds, such as benzene rings. The electrophile attacks the aromatic ring, leading to the substitution of one atom or group with the electrophile.

An example of an electrophilic substitution reaction is the nitration of benzene, where a nitro group (-NO2) replaces a hydrogen atom on the benzene ring.

Electrophilic Substitution Examples

Aromatic Substitution (e.g., Friedel-Crafts Alkylation)

Reaction: $R–H+E^{+} \rightarrow R–E + H^{+}$

Example: $C^{6}H^{6} + CH_3Cl + AlCl_3 \rightarrow C_6H_5CH_3 + HCl$

Halogenation of Alkanes

Reaction: $R–H + X_2 \rightarrow R–X + HX$

Example: $CH_4 + Cl_2 \rightarrow CH_3Cl + HCl$

Electrophilic Substitution Reactions of Benzene

Electrophilic substitution reactions of benzene are a class of reactions where an electrophile (electron-deficient species) replaces one of the hydrogen atoms in a benzene ring. Benzene is a highly stable and aromatic compound, and electrophilic substitution reactions allow for the introduction of various functional groups onto the benzene ring.

Some common electrophilic substitution reactions of benzene include: • Nitration of Benzene

- Sulphonation of Benzene
- Halogenation of Benzene
- Friedel-Crafts Alkylation
- Friedel-Crafts Acylation

Free Radical Substitution Reactions

Free radical substitution reactions are a type of chemical reaction in which a radical (a species with an unpaired electron) replaces another atom or group of atoms in a molecule. Free radical substitution reactions can be described as a three-step process:

- Initiation
- Propagation

- Termination

Examples of Free Radical Substitution

Some examples of free radical substitution are listed as follows: • Halogenation of Alkanes

- Chlorination of Methane
- Bromination of Ethane
- Idonination of Propane

Organometallic Substitution Reactions

Organometallic substitution reactions involve the replacement of one or more ligands (organic groups or other atoms) in an organometallic compound with different ligands.

Organometallic Substitution Examples

Organometallic substitution reactions can be broadly categorized into several types, including ligand substitution, oxidative addition, and reductive elimination. Here are some examples:

Ligand Substitution

Ligand substitution reactions involve the exchange of one ligand for another. Commonly, a new ligand displaces an existing ligand from the coordination sphere of the metal center. For example, in the reaction of ferrocene (an organometallic compound) with iodine:

$$Fe(C_5H_5)_2 + I2 \rightarrow Fe(C_5H_5)I + C_5H_5I$$

Here, iodine (I2) substitutes for one of the cyclopentadienyl (C5H5) ligands.

Oxidative Addition

Oxidative addition reactions typically involve the addition of a new ligand and the simultaneous increase in the oxidation state of the metal center. For example, in the reaction of a palladium complex with hydrogen gas:

$$Pd(PPh3)4 + H_2 \rightarrow Pd(PPh3)2H2 + 2PPh3$$

In this reaction, hydrogen gas (H2) undergoes oxidative addition to palladium (Pd) to form a dihydride complex.

Reductive Elimination

Reductive elimination reactions involve the removal of one or more ligands from the metal center and the simultaneous decrease in the oxidation state of the metal. For example, in the reductive elimination of ethylene from a platinum complex:

$$Pt(C2H_4)_2 + Cl_2 \rightarrow PtCl_2 + C2H_4$$

In this reaction, the platinum complex loses ethylene (C2H4) to form platinum(II) chloride.

Substitution with a Nucleophile (Reaction with RMgX)

Organometallic compounds can also undergo nucleophilic substitution reactions, where a nucleophile replaces one of the ligands. For instance, in the reaction of a Grignard reagent (nucleophile) with a transition metal complex:

$$M(CO)_6 + RMgX \rightarrow M(CO)5R + MgXCl$$

ADDITIONANDELIMINATIONREACTIONSINALKENESANDALKYNES ADDITION REACTIONS OF ALKENES AND ALKYNES

Addition reactions are, essentially, the opposite of elimination reactions. Recall that an elimination reaction had the following general form:

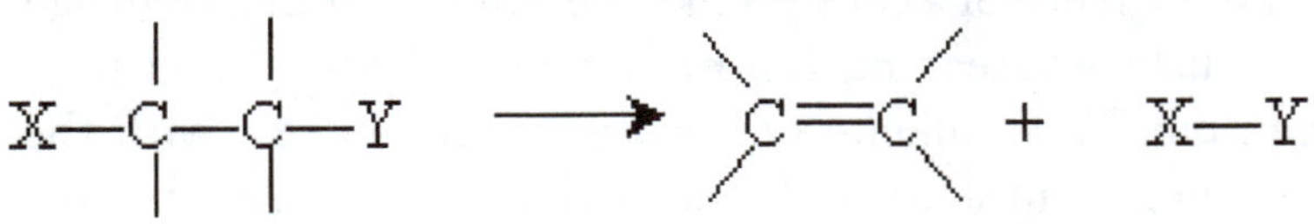

An addition reaction is simply the reverse of this process:

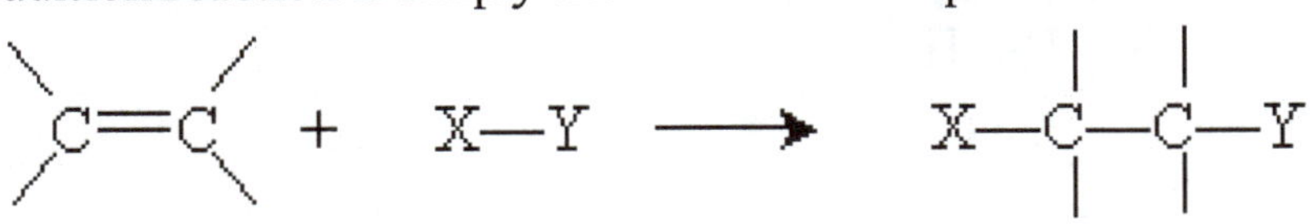

Clearly, the same substances are present in both cases, so the difference between the reaction that prefers one side (say, the products) over the other side (the reactants) is the conditions under which the reaction occurs: other chemicals that may be present, temperature, and so on.

For alkynes, the general elimination reaction is shown below, followed by the corresponding addition reaction.

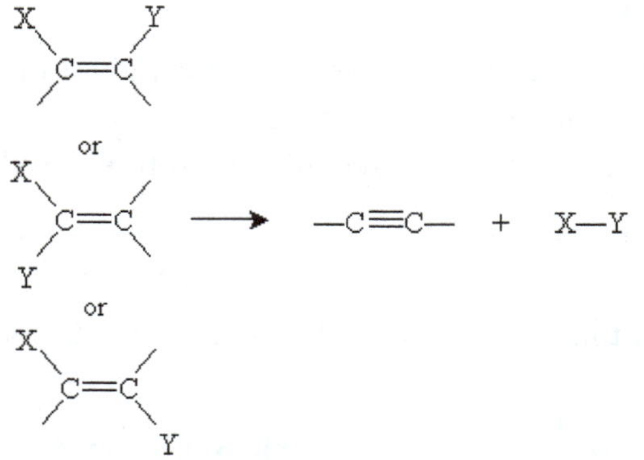

Hydrogenation

Alkenes and alkynes are called unsaturated hydrocarbons because, as the name indicates, the carbon atoms are not "saturated" with hydrogens, owing to the presence of double or triple carbon-carbon bonds. As a result, these substances tend to be more reactive at the site of the multiple bond than a saturated hydrocarbon (an alkane or cycloalkane), which cannot accept any additional atoms. In the presence of heat, saturated hydrocarbons can eliminate hydrogen gas to form unsaturated hydrocarbons (alkenes and alkynes). The reverse process, hydrogenation, can be performed in the presence of a catalyst like metallic platinum. Hydrogen gas reacts with the metal surface, breaking the hydrogen-hydrogen bond to form weaker metal-hydrogen bonds. An alkene or alkyne can then react with the metal in a similar manner, then form stronger bonds with two or more hydrogen atoms. This process is illustrated below for a generic alkene.

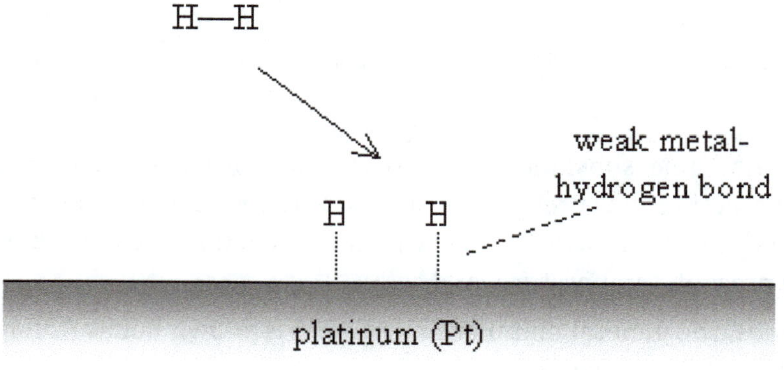

The alkene forms similar weak carbon-metal bonds at the site of the double bond.

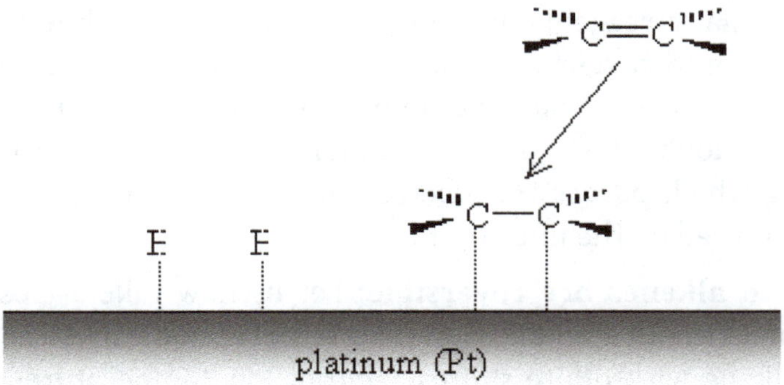

Reactions of Alkenes: Electrophilic Addition

The double-bonded carbons of an alkene are electron-rich, that is, the electron density is high in the region of the double bond. Therefore, the "signature" reaction of alkenes involves initial attack on an electrophile. Instead of a substitution, alkenes undergo electrophilic addition, a reaction in which a two-component reactant adds across the double bond. The reaction begins with an electrophilic attack by the double bond onto the reactant which produces a carbocation that then undergoes nucleophilic attack. In the case of unsymmetrical alkenes (where the groups attached to the double-bonded carbons are not exactly the same), the most stable carbocation is produced. This reaction is regioselective, that is, we can predict the orientation of reactant addition across the double bond. If we designate the reagent as E (for electrophile) or N (for nucleophile), the reaction would proceed as outlined below.

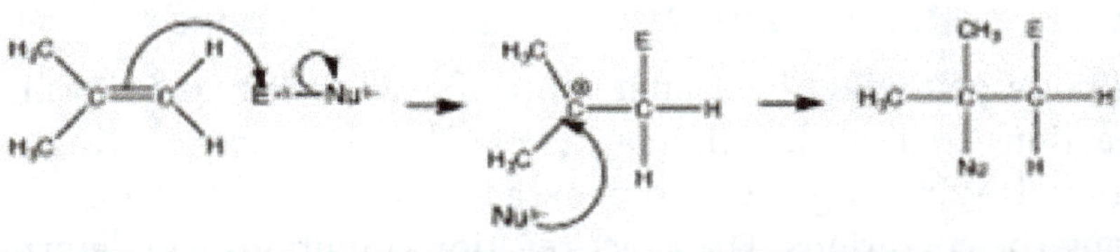

Reagent	Electrophile	Nucleophile	Typical Conditions
HBr	H^+	Br^-	Low Temp
H_2O	H^+	H_2O (with loss of H+ after addition)	Aqueous acid/Low Temp
ROH	H^+	ROH	ROH/H_3O^+
Br_2	Br^+	Br^-	Br_2/CCl_4
BrOH	Br^-	^-OH	Br_2/H_2O
BrOR	Br^+	^-OR	Br_2/ROH

The intermediate carbocation is the tertiary carbocation, (rather than the primary carbocation that would be produced by addition to the =CH2 end of the double bond). This pattern of reaction is referred to as Markovinkov addition, after the person who first discovered that HBr adds in this way to a double bond.

We can classify many reagents as combinations of electrophile and nucleophile and, in this way, predict how they will add across the double bond. Examples of such reagents are shown (). Rather than memorizing the product of every type of addition across a double bond, it is much more productive to write a mechanism by determining which part is the electrophile, adding it to give the most stable carbocation, followed by the nucleophile.

Additions to alkenes are reversible: Let us now take a closer look at the addition of water across a double bond. Such a reaction can be accomplished by reacting the alkene with dilute sulfuric acid at low temperatures. The first step is addition of a proton to produce the most stable carbocation—which is then attacked by water (the nucleophile). The final product is the alcohol that forms after a proton is transferred to water. In this case, we can consider the proton (or more accurately H3O+) as a catalyst since it is regenerated at the end of the reaction sequence.

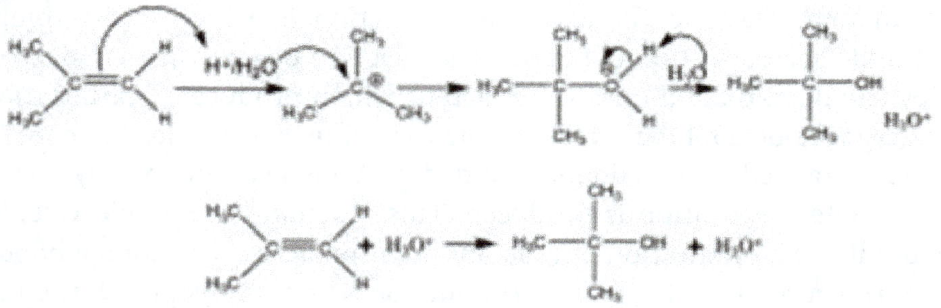

Acid-catalyzed addition of water across a double bond

At this point you might be asking yourself: well didn't we just talk about the reverse reaction—that is, the elimination of H2O from alcohols to give alkenes? Indeed we did! Many organic reactions are reversible, it is just a matter of manipulating the conditions. The exact reaction conditions will determine which reaction is favored.

As is the case with most addition reactions, the addition of water across an alkene is exothermic, that is, DH is negative because stronger (sigma) bonds are formed during the reaction and energy is released into the environment. A typical energy diagram is shown below (Fig. 8.7).

This means that DH for the elimination reaction must be positive (i.e. going from right to left on the diagram above). The question then is: why does an elimination reaction ever occur? To answer that, we have to recall that the thermodynamic criterion for a reaction to proceed is not simply a negative enthalpy change, but rather a negative change in the Gibbs change (ΔG). Recall that $\Delta G = \Delta H - T\Delta S$.

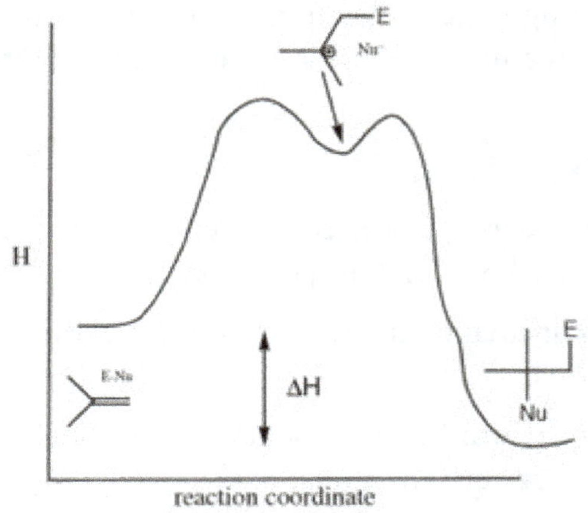

Fig. 8.7 Reaction energy diagram for addition/elimination across a double bond

The change in entropy also influences the thermodynamic favorability of a reaction. In the elimination reaction, two molecules (alkene and water) are produced from one alcohol molecule; the entropy change will be positive. (Recall that entropy is associated with the number of possible arrangements of the system.

Since two molecules will have more possible arrangements than one, this reaction will always be accompanied by an increase in entropy of the system.) To encourage the equilibrium to shift to the right (the addition reaction) we need increase the temperature, which will increase the magnitude of the $-T\Delta S$ term, making ΔG more negative (assuming that ΔS is positive). In general, additions to double bonds are carried out at lower temperatures, while elimination reactions involve heating the reaction solution. Another way to influence the equilibrium state is to change the relative concentrations of reactants or products.

REARRANGEMENT REACTIONS AND CARBOCATION STABILITY INTRODUCTION TO CARBOCATION

A carbocation is a cation with a positively charged carbon atom. Some examples are methenium CH_3^+, methanium CH_5^+, and vinyl $C_2H_3^+$ cations. Carbocations containing more than one positively charged carbon can be found very rarely. For example, ethylene dication $C_2H_4^{2+}$. According to the IUPAC, a carbocation is defined as any cation that contains an even number of electrons with a significant portion of the positive charge residing on a carbon atom. The carbon atom of a carbocation is sp2 hybridised. The three hybridised orbitals of the carbon atom are used for single bonding to three substituents, and the remaining p-orbital is empty.

Carbocations are important reactive intermediates in many organic synthesis reactions. Carbocation rearrangements are generally reactions in organic Chemistry where the movement of a carbocation from an unstable state to a more stable state occurs through the molecular rearrangement in the structure. After the carbocation rearrangement, the new molecule is the structural isomer of the initial molecule.

Classification of Carbocations

- **Primary Carbocation:** Here, the carbon having the positive charge is attached to one carbon of an alkyl or aryl group.
- **Secondary Carbocation:** In this carbon, a positive charge is attached to two other carbons.
- **Tertiary Carbocation:** In this case, the carbon having the positive charge is attached to three carbon atoms.

In methyl carbocation, the carbon bearing a positive charge is attached to only three hydrogen atoms.

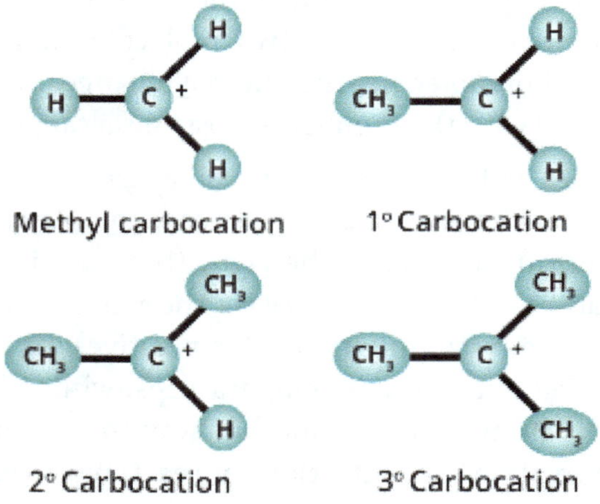

Fig. 8.8. Type of Carbocation

Generation of Carbocation

The different methods available to generate stable or unstable carbocation are:

1. Ionisation: When direct ionisation of an alkyl or aryl compound occurs, a leaving group attached to a carbon atom leaves with its pair of electrons, forming a carbocation. Stable carbocations can be produced in super acids. For example, the addition of alkyl fluorides to SbF5 forms a stable carbonation.

RF+SbF5 → R++SbF-6RF+SbF5 → R++SbF6-

2. Ionisation After an Initial Reaction: Occasionally, the functional groups in an organic compound may undergo an initial reaction first. During this reaction, the functional group converts to a good leaving group, and the intermediate formed then ionises into a carbocation after the removal of the leaving group. Protonation of organic alcohol generates an oxonium ion which ionises to a carbocation by losing a water molecule.

$$R\text{-}OH \rightarrow H{+}R\text{-}OH2{+} \rightarrow R{+}{+}H_2OR\text{-}OH \rightarrow H{+}R\text{-}OH_2{+} \rightarrow R{+}{+}H_2O$$

3. Attack of p-system on Electrophiles: When an electrophile attacks one of the atoms of a p-system in an organic molecule, the adjacent atom acquires a positive charge. In such cases, if the positive atom is carbon, then a carbocation is formed. When the p-system of an alkene or alkyne gets attacked with an electrophile like proton, bromonium ion, etc., a positive charge develops on the adjacent carbon and leads to the formation of a carbocation intermediate. For example,

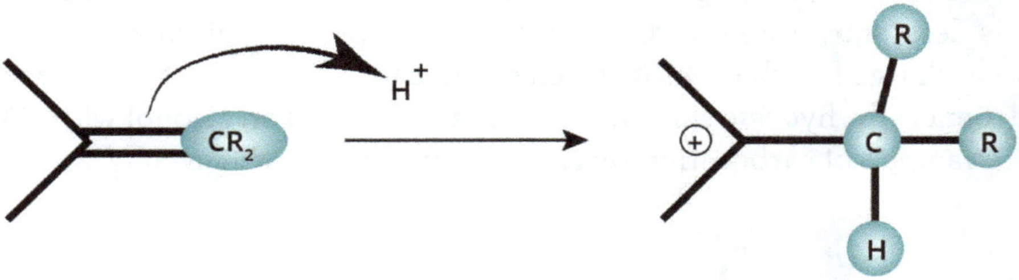

Fig. 8.9 Attack of p-system in Alkene on an Electrophile.

Stability of Carbocation

The stability of a carbocation is decided by three main factors, inductive effect, hyperconjugation, or resonance effect. Carbocations are the most stable when the charge is on a tertiary carbon and the least stable on a primary carbon. The order of stability is 3o carbocation > 2occarbocation > 1o carbocation > Methyl carbocation.

1. Inductive Effect: If the groups attached to positively charged carbon are electron releasing, they will delocalise the positive charge on carbon and will decrease the intensity of the positive charge on carbon. This stabilises the carbocation. The more alkyl groups are attached, the more stable the carbocation will be as alkyl groups release electrons.

2. Hyperconjugation: If in any carbocation, the sigma bond in conjugation with vacant p orbital on carbon has a positive charge, then it participates in delocalisation. Thus, they spread the charge all over the alkyl groups. This effect is known as hyperconjugation or no bond resonance. The structure of the ethyl carbocation has the following series of resonance forms, according to the theory of hyperconjugation.

3. Resonance: Conjugation with multiple bonds or lone pairs of electrons increases the stability of a carbocation. For example, triphenylmethyl cation has higher stability than diphenylmethyl cation. In the case of triphenylmethyl cation, delocalisation of positive charge is more as it has more resonating structures.

Rearrangement of Carbocation

Carbocation rearrangements are the movement of a carbocation from an unstable state to a more stable state through structural rearrangement to form a structural isomer of the initial molecule. Rearrangements of carbocation are of two types: hydride shift and alkyl shift. When rearranged, the molecules can also undergo unimolecular substitution (SN1) or unimolecular elimination (E1) reactions.

Hydride Shift

In hydride shift, hydrogen moves over from one carbon to a neighbouring, which is less substituted carbon. Here, the hydrogen migrates towards the positively charged carbon with its electron pair, leaving a stable carbocation behind. Generally, hydride shift occurs in the reaction of an alcohol with HBr, HCl, and HI. Examples of carbocation rearrangement are given below (Fig. 8.10).

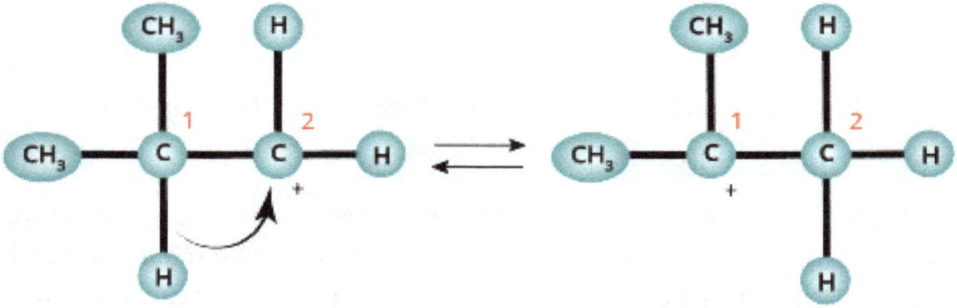

Fig. 8.10 1-2 Hydride Shift

Below is a carbocation rearrangement example, which is a reaction between an alcohol and hydrogen chloride.

- **Protonation and Loss of Water Molecule:** In this SN1 reaction, -OH which leaves the organic alcohol forms a carbocation on carbon (C-3) after receiving a proton from the nucleophile to produce an alkyloxonium ion.
- **Hydride Shift:** The hydrogen atom attached to the carbon atom adjacent to the original carbon (C-2) undergoes a hydride shift in this case.
- **Attack of Cl Atom:** The Cl atom can now attack the carbocation and form a more stable structure because of hyperconjugation. The carbocation here is the rearranged most stable carbocation, which is tertiary carbon.

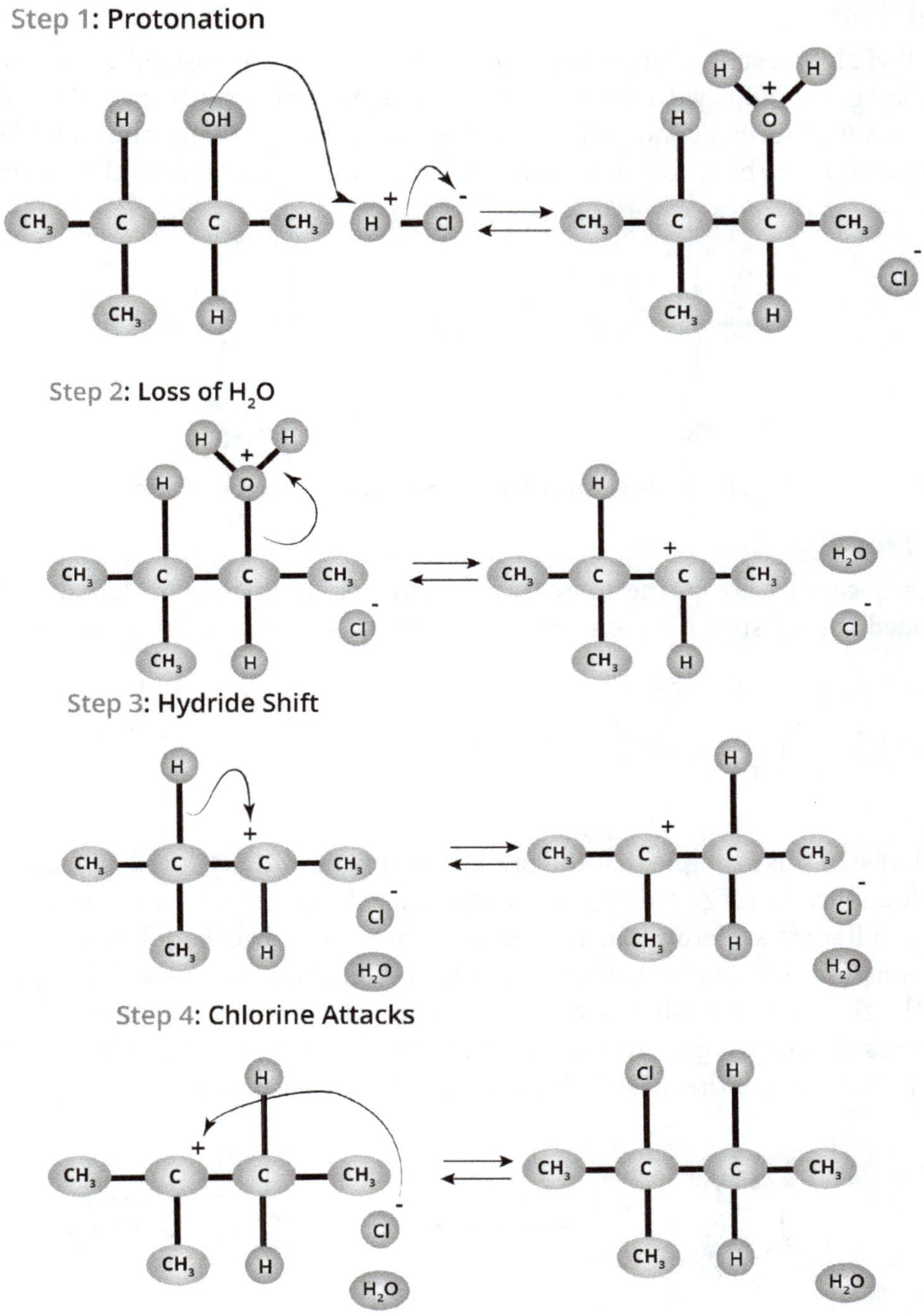

Fig. 8.11 Hydride Shift Rearrangement of Carbocation

Alkyl Shift

Alkyl Shift reactions are similar to hydride shifts. Here, instead of the proton, the alkyl group shifts with the nucleophile. The group that shifts takes its electron pair with it to form a bond with the nearby carbocation. Finally, the shifted alkyl group and the carbocation with positive charge switch positions on the molecule.

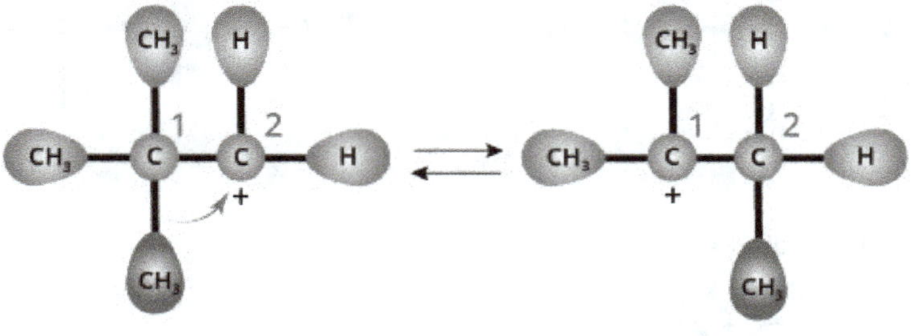

Fig. 8.12 Alkyl Shift Rearrangement of Carbocation

Rearrangements

Carbocations are prone to rearrangement via 1,2-hyride or 1,2-alkyl shifts provided at each step, a more stable carbocation is generated. For example:

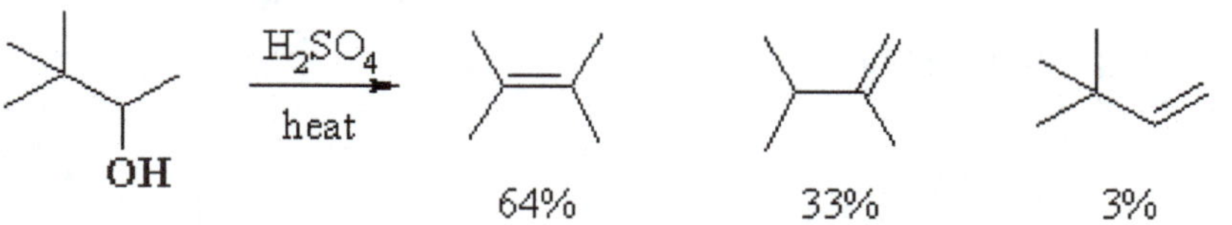

Notice in this example that the product we might initially "predict" based on a simple application of Zaitsev's rule, is only formed in 3% yield, and instead, products with a different skeleton (i.e. a different hydrocarbon backbone) dominate. The reaction proceeds via protonation to give the better leaving group which departs to give the 2o carbocation shown. A methyl group rapidly migrates taking its bonding electrons along, giving a new skeleton and a more stable 3o carbocation which can then lose H+ to give the more stable alkene as the major product.

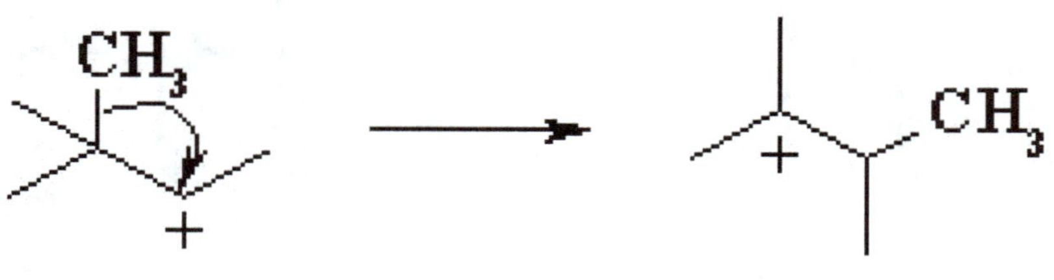

2° carbocation to 3° carbocation

This is an example of a 1,2-alkyl shift. The numbers indicate that the alkyl group moves to an adjacent position.

Similar migrations of H atoms, 1,2-hydride shifts are also known.

A good rule of thumb for carbocations....."If they can rearrange, they will" (that means if a more stable carbocation can be formed, then expect them to rearrange.

Rearrangments of simple alkyl carbocations are of the 1,2-type (this means that only a group on an adjacent position migrates).

As a potential sign of rearrangement, look for carbocations adjacent to branch points (see the example above). Carbocations in the middle of unbranched chains will not usually rearrange.

If a product seems to have a different hydrocarbon skeleton, that is an indicator to check for a carbocation rearrangement via alkyl shifts (see above).

If a product seems to have a new functional group in a new location, that is an indicator to check for a carbocation rearrangement via a hydride shift.

Carbocation

Product

Reaction with nucleophile

They can combine with nucleophiles by accepting an electron pair. If all three groups on carbocation are different racemic mixture can be obtained.

Addition reaction

They add to the double bond to create a new positively charged center.

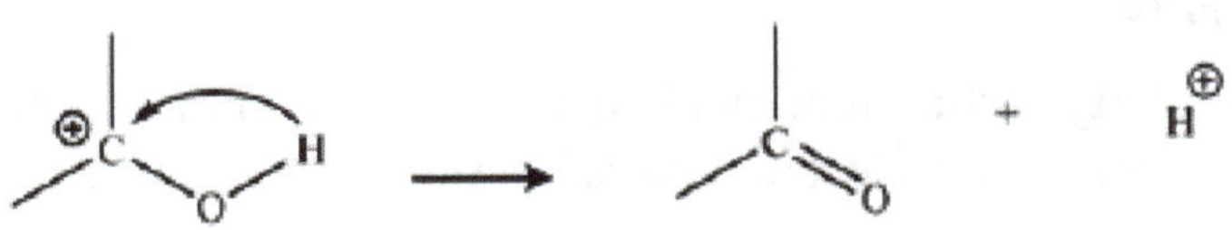

Proton removal

They may result in the removal of a proton from an adjacent carbon atom resulting in the formation of the double bond.

Factors affecting the stability of carbocation

Their stability is influenced by various factors. They are:

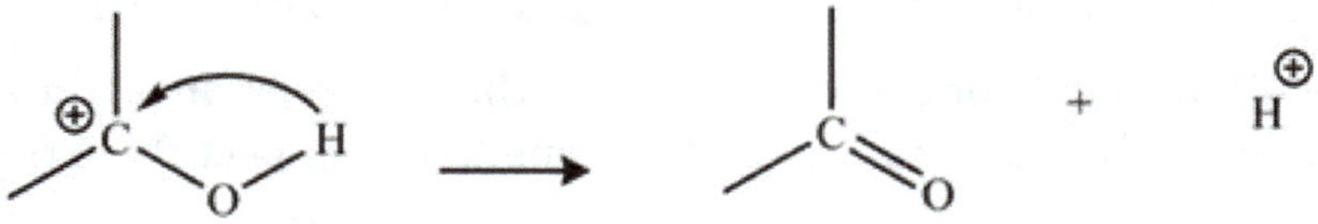

Resonance

The carbocation stability rises as the number of resonances increases. Furthermore, the more resonating structures there are, the more stable the carbocations are. The delocalization of the positive charge reduces electron loss even further, increasing total stability. In comparison, it has been discovered that the resonance effect has a significantly bigger influence than replacement. Structures with resonance are more stable than others.

Inductive effect

Because the alkyl group has an electron-donating effect (+I), the stability of the carbocation is enhanced as the attached group's donating ability grows.

Alkyl groups can donate more electrons inductively than hydrogen groups because they are bigger, more polarizable, and have more bonding electrons. Thus, the more alkyl substitutions there are, the more stable the carbocation.

Hyperconjugation effect

Hyperconjugation is the process of donating electrons from the parallel overlap of p orbitals with neighboring hybridized orbitals from sigma bonds. As a result, the total amount of carbon atom groups connected to the positively charged carbon determines the overall carbocation stability.

With the increase in substitution, hyperconjugation is also increased, which increases stability. The more hyperconjugation there is the more stable the system.

Electronegativity

Electronegativity is the capacity of an atom to attract or pull electrons to itself. Thus, the higher an atom's electronegativity, the more electrons it attracts.

The stability is also directly controlled by the electronegativity of a positively charged carbon atom. When the electronegativity of the carbon atom increases, the stability of the carbocation decreases, and vice versa.

As an example, sp > sp2 > sp3. In the case of vinylic carbocation, the hybridization of the positive carbon atom is sp, with a stronger electronegativity than the sp2 alkyl carbocation hybridized carbon. As a result, the stability of the primary vinylic carbocation is lower than that of the primary alkyl carbocation.

Steric effect

Because alkyl carbocations are predominantly produced from tetrahedral alkyl halides, a link between steric relief and carbocation formation can be established. In such circumstances, the carbon-carbon bond angles change from 1090282 to 1200 during the production of carbocations. As a result, carbon with bulky groups around it is expected to benefit more from this carbocationic conversion. The stability of carbocation increases from left to right in the figure below:

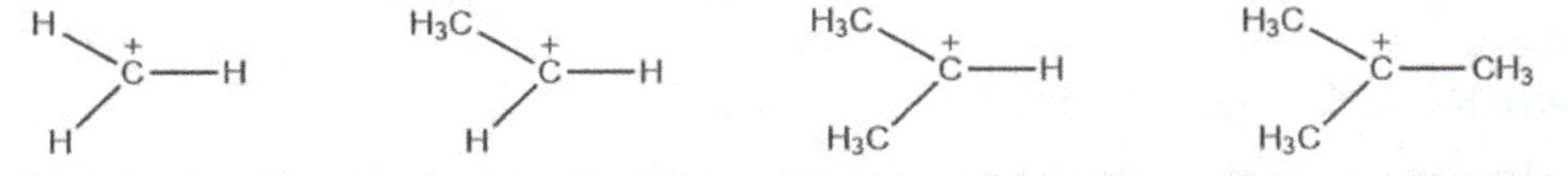

Methyl carbocation Primary carbocation Secondary carbocation Tertiary carbocation

RADICAL REACTIONS AND POLYMERIZATION

Radical Polymerization of Alkenes, Polymers

Prior to the early 1920's, chemists doubted the existence of molecules having molecular weights greater than a few thousand. This limiting view was challenged by Hermann Staudinger, a German chemist with experience in studying natural compounds such as rubber and cellulose. In contrast to the prevailing rationalization of these substances as aggregates of small molecules, Staudinger proposed they were made up of macromolecules composed of 10,000 or more atoms. He formulated a polymeric structure for rubber,based on a repeating isoprene unit (referred to as a monomer). For his contributions to chemistry, Staudinger received the 1953 Nobel Prize. The terms polymer and monomer were derived from the Greek roots poly (many), mono (one) and meros (part).

Recognition that polymeric macromolecules make up many important natural materials was followed by the creation of synthetic analogs having a variety of properties. Indeed, applications of these materials as fibers, flexible films, adhesives, resistant paints and tough but light solids have transformed modern society. Some important examples of these substances are discussed in the following sections. There are two general types of polymerization reactions: addition polymerization and condensation polymerization. In addition polymerization, the monomers add to one another in such a way that the polymer contains all the atoms of the starting monomers. Ethylene molecules are joined together in long chains.Strictly speaking, this chapter is on radical reactions, which include most addition polymerization processes. However, in order to cover the topic of polymers properly, we are including a brief section on condensation polymerization (usually based on substitution chemistry) here as well.

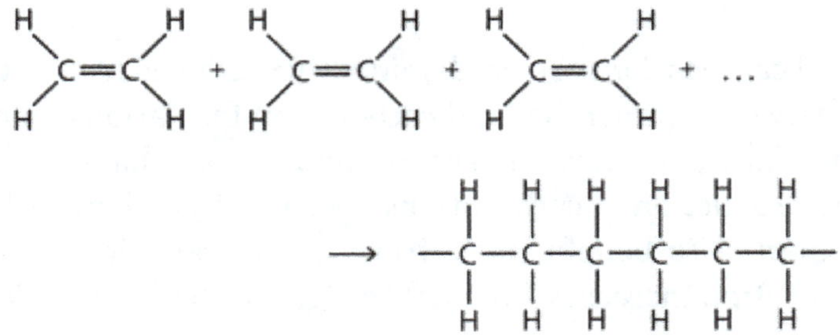

Chain-Reaction (Addition) Polymerization

The polymerization can be represented by the reaction of a few monomer units:

The bond lines extending at the ends in the formula of the product indicate that the structure extends for many units in each direction. Notice that all the atoms—two carbon atoms and four hydrogen atoms—of each monomer molecule are incorporated into the polymer structure. Because displays such as the one above are cumbersome, the polymerization is often abbreviated as follows:

$$nCH_2=CH_2 \rightarrow [\ CH_2CH_2\ ]_n$$

During the polymeriation of ethene, thousands of ethene molecules join together to make poly(ethene) - commonly called polythene. The reaction is done at high pressures in the presence of a trace of oxygen as an initiator.

$$nCH_2=CH_2 \rightarrow [-CH_2CH_2-]_n$$

Note that all the monomers have carbon-to-carbon double bonds. Many polymers are mundane (e.g., plastic bags, food wrap, toys, and tableware), but there are also polymers that conduct electricity, have amazing adhesive properties, or are stronger than steel but much lighter in weight.

Step 1: Chain Initiation

The oxygen reacts with some of the ethene to give an organic peroxide. Organic peroxides are very reactive molecules containing oxygen-oxygen single bonds which are quite weak and which break easily to give free radicals. You can short-cut the process by adding other organic peroxides directly to the ethene instead of using oxygen if you want to. The type of the free radicals that start the reaction off vary depending on their source. For simplicity we give them a general formula: Ra.

Step 2: Chain Propagation

In an ethene molecule, CH2=CH2, the two pairs of electrons which make up the double bond aren't the same. One pair is held securely on the line between the two carbon nuclei in a sigma bond. The other pair is more loosely held in an orbital above and below the plane of the molecule in a p bond.

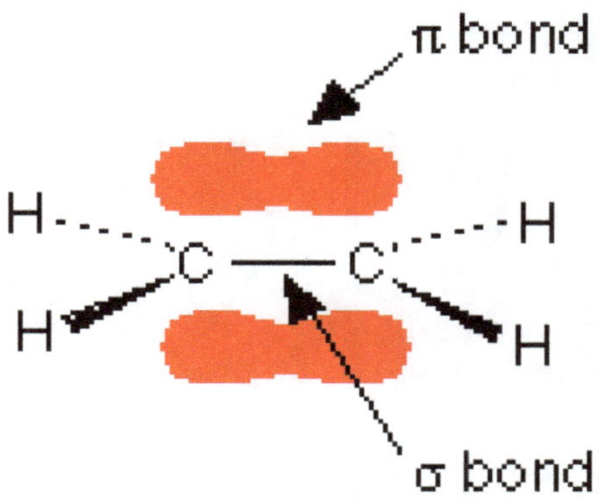

Imagine what happens if a free radical approaches the p bond in ethene.

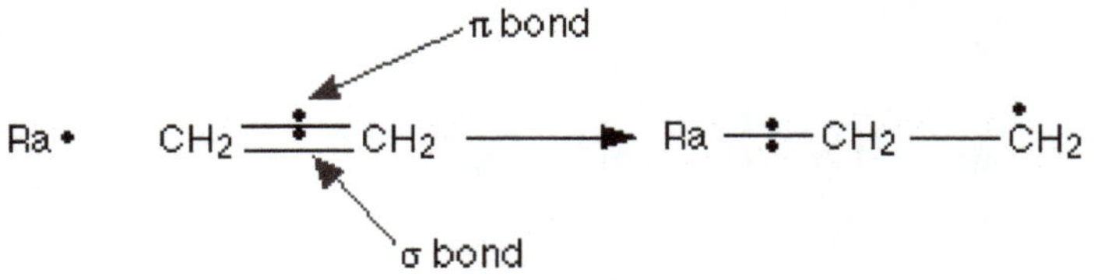

The sigma bond between the carbon atoms isn't affected by any of this. The free radical, Ra, uses one of the electrons in the p bond to help to form a new bond between itself and the left hand carbon atom – a radical addition step. The other electron returns to the right hand carbon.

9

REACTION DYNAMICS IN COMPLEX SYSTEMS

High-quality and quantitative in situ experiments are needed in order to advance the field. Towards this overarching goal, our lab has been developing new instruments and theoretical methods that are seeing applications in such areas as physical materials, chemical and materials biology, and biophysics. Current research topics include:

- The dynamics of single protein molecules, the fundamental building blocks of life;
- Instrumentation to capture the real-time motion and dynamics of single molecules/ particles as they explore complex, 3D environments;
- Design and characterization of functional nanostructures, as well as investigation of their interaction with biological systems;
- Using light to control artificial swimmers and to develop new types of active matter;
- New theoretical frameworks and statistical learning techniques to afford an objective and quantitative understanding of single-molecule and single-particle measurements.

REACTION NETWORKS AND METABOLIC PATHWAYS

In biochemistry, a metabolic pathway is a linked series of chemical reactions occurring within a cell. The reactants, products, and intermediates of an enzymatic reaction are known as metabolites, which are modified by a sequence of chemical reactions catalyzed by enzymes.In most cases of a metabolic pathway, the product of one enzyme acts as the substrate for the next. However, side products are considered waste and removed from the cell.

Different metabolic pathways function in the position within a eukaryotic cell and the significance of the pathway in the given compartment of the cell. For instance, the electron transport chain and oxidative phosphorylation all take place in the mitochondrial membrane.In contrast, glycolysis, pentose phosphate pathway, and fatty acid biosynthesis all occur in the cytosol of a cell.

There are two types of metabolic pathways that are characterized by their ability to either synthesize molecules with the utilization of energy (anabolic pathway), or break down complex molecules and release energy in the process (catabolic pathway).

The two pathways complement each other in that the energy released from one is used up by the other. The degradative process of a catabolic pathway provides the energy required to conduct the biosynthesis of an anabolic pathway. In addition to the two distinct metabolic pathways is the amphibolic pathway, which can be either catabolic or anabolic based on the need for or the availability of energy.

Pathways are required for the maintenance of homeostasis within an organism and the flux of metabolites through a pathway is regulated depending on the needs of the cell and the availability of the substrate. The end product of a pathway may be used immediately, initiate another metabolic pathway or be stored for later use. The metabolism of a cell consists of an elaborate network of interconnected pathways that enable the synthesis and breakdown of molecules (anabolism and catabolism).

Each metabolic pathway consists of a series of biochemical reactions that are connected by their intermediates: the products of one reaction are the substrates for subsequent reactions, and so on. Metabolic pathways are often considered to flow in one direction. Although all chemical reactions are technically reversible, conditions in the cell are often such that it is thermodynamically more favorable for flux to proceed in one direction of a reaction. For example, one pathway may be responsible for the synthesis of a particular amino acid, but the breakdown of that amino acid may occur via a separate and distinct pathway. One example of an exception to this "rule" is the metabolism of glucose. Glycolysis results in the breakdown of glucose, but several reactions in the glycolysis pathway are reversible and participate in the re-synthesis of glucose (gluconeogenesis).

Glycolysis was the first metabolic pathway discovered:

- As glucose enters a cell, it is immediately phosphorylated by ATP to glucose 6-phosphate in the irreversible first step.
- In times of excess lipid or protein energy sources, certain reactions in the glycolysis pathway may run in reverse to produce glucose 6-phosphate, which is then used for storage as glycogen or starch.

- Metabolic pathways are often regulated by feedback inhibition.
- Some metabolic pathways flow in a 'cycle' wherein each component of the cycle is a substrate for the subsequent reaction in the cycle, such as in the Krebs Cycle (see below).
- Anabolic and catabolic pathways in eukaryotes often occur independently of each other, separated either physically by compartmentalization within organelles or separated biochemically by the requirement of different enzymes and co factors.

Metabolic Pathways

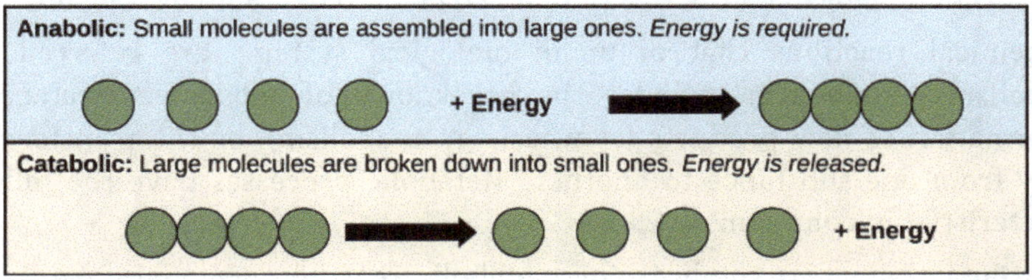

Fig. 9.1 Anabolic pathway: small molecules are assembled into larger ones. Energy is typically required. Catabolic pathway: large molecules are broken down into small ones. Energy is typically released.

Metabolic pathways can be broadly divided into two categories based on their effects. Photosynthesis, which builds sugars out of smaller molecules, is a "building up," or anabolic, pathway. In contrast, cellular respiration breaks sugar down into smaller molecules and is a "breaking down," or catabolic, pathway.

Anabolic pathways build complex molecules from simpler ones and typically need an input of energy. Building glucose from carbon dioxide is one example. Other examples include the synthesis of proteins from amino acids, or of DNA strands from nucleic acid building blocks (nucleotides). These biosynthetic processes are critical to the life of the cell, take place constantly, and use energy carried by ATP and other short-term energy storage molecules.

Catabolic pathways involve the breakdown of complex molecules into simpler ones and typically release energy. Energy stored in the bonds of complex molecules, such as glucose and fats, is released in catabolic pathways. It's then harvested in forms that can power the work of the cell (for instance, through the synthesis of ATP).

One final but important note: the chemical reactions in metabolic pathways don't take place automatically, without guidance. Instead, each reaction step in a pathway is facilitated, or catalyzed, by a protein called an enzyme. You can learn more about enzymes and how they control biochemical reactions in the enzymes topic.

Chemical Reactions in Metabolic Processes

In order for a chemical reaction to take place, the reacting molecules (or atoms) must first collide and then have sufficient energy (activation energy) to trigger the formation of new bonds. Although many reactions can occur spontaneously, the presence of a catalyst accelerates the rate of the reaction because it lowers the activation energy required for the reaction to take place. A catalyst is any substance that accelerates a reaction but does not undergo a chemical change itself. Since the catalyst is not changed by the reaction, it can be used over and over again.

Chemical reactions that occur in biological systems are referred to as metabolism. Metabolism includes the breakdown of substances (catabolism), the formation of new products (synthesis or anabolism), or the transferring of energy from one substance to another. Metabolic processes have the following characteristics in common:

- Enzymes act as catalysts for metabolic reactions. Enzymes are proteins that are specific for particular reactions. The standard suffix for enzymes is "ase," so it is easy to identify enzymes that use this ending (although some do not). The substance on which the enzyme acts is called the substrate. For example, the enzyme amylase catalyzes the breakdown of the substrate amylose (starch) to produce the product glucose.The induced fit model describes how enzymes work. Within the protein (the enzyme), there is an active site with which the reactants readily interact because of the shape, polarity, or other characteristics of the active site. The interaction of the reactants (substrate) and the enzyme causes the enzyme to change shape. The new position places the substrate molecules in a position favorable to their reaction and accelerates the formation of the product.
- Adenosine triphosphate (ATP) is a common source of activation energy for metabolic reactions. In Figure 1, the wavy lines between the last two phosphate groups of the ATP molecule indicate high energy bonds. When ATP supplies energy to a reaction, it is usually the energy in the last bond that is delivered to the reaction. In the process of giving up this energy, the last phosphate bond is broken and the ATP molecule is converted to ADP (adenosine diphosphate) and a phosphate group (indicated by Pi). In contrast, new ATP molecules are assembled by phosphorylation when ADP combines with a phosphate group using energy obtained from some energy rich molecule (like glucose).

- Cofactors are nonprotein molecules that assist enzymes. A holoenzyme is the union of the cofactor and the enzyme (called an apoenzyme when part of a holoenzyme). If cofactors are organic, they are called coenzymes and usually function to donate or accept some component of a reaction, often electrons. Some vitamins are coenzymes or components of coenzymes. Inorganic cofactors are often metal ions, such as Fe++.

CHEMICAL REACTIVITY IN BIOLOGICAL SYSTEMS

Within biological systems there are six major classes of biochemical reactions that are mediated by enzymes. These include group transfer reactions, the formation/removal of carbon-carbon double bonds, isomerization reactions, ligation reactions, hydrolysis reactions, and oxidation-reduction reactions.

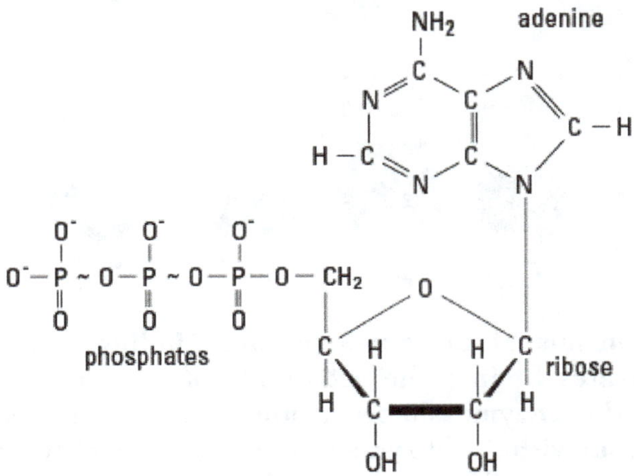

Adenosine Triphosphate (ATP)

Fig. 9.2. The high energy bonds of adenosine triphosphate (ATP).

The chemical reactions of metabolism are organized into metabolic pathways, in which one chemical is transformed through a series of steps into another chemical, each step being facilitated by a specific enzyme. Enzymes are crucial to metabolism because enzymes act as catalysts – they allow a reaction to proceed more rapidly. In addition, enzymes can provide a mechanism for cells to regulate the rate of a metabolic reaction in response to changes in the cell's environment or to signals from other cells, through the activation or inhibition of the enzymes activity. Enzymes can also allow organisms to drive desirable reactions that require energy that will not occur by themselves, by coupling them to spontaneous reactions that release energy. Enzyme shape is critical to the function of the enzyme as it determines the specific binding of a reactant. This can occur by a lock and key model where the reactant is the exact shape of the enzyme binding site, or by an induced fit model, where the contact of the reactant with the protein causes the shape of the protein to change in order to bind to the reactant.

Group Transfer Reactions

In group transfer reactions, a functional group will be transferred from one molecule that serves as the donor molecule to another molecule that will be the acceptor molecule. The transfer of an amine functional group from one molecule to another is common example of this type of reaction and is shown in Figure 9.4.

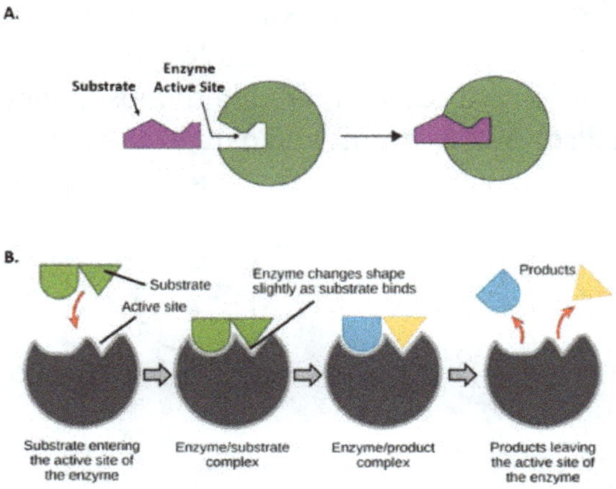

Figure 9.3 Mechanisms of Enzyme-Substrate Binding. (A) In the Lock and Key Model, substrates fit into the active site of the enzyme with no further modifications to the enzyme shape required. (B) In the Induced Fit Model, substrate interaction with the enzyme causes the shape of the enzyme to change to better fit the substrate and mediate the chemical reaction.

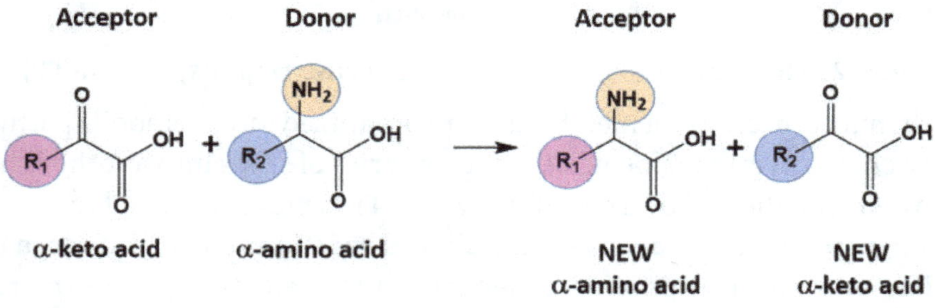

Fig. 9.4 Transfer of an Amine Functional Group. A common group transfer reaction in biological systems is one that is used to produce a-amino acids that can then be used for protein synthesis. In this reaction, one a-amino acid serves as the donor molecule and an a-keto acid (these molecules contain a carboxylic acid functional group and a ketone functional group separated by one a-carbon) serves as the acceptor. In the acceptor molecule, the carbonyl oxygen is replaced with the amine functional group, whereas in the donor molecule, the amine functional group is replaced by an oxygen forming a new ketone functional group.

Isomerization Reactions

In isomerization reactions a single molecule is rearranged such that it retains the same molecular formula but now has a different bonding order of the atoms forming a structural or stereoisomer. The conversion of glucose 6-phosphate to fructose 6-phosphate is a good example of an isomerization reaction and is shown in Fig. 9.5.

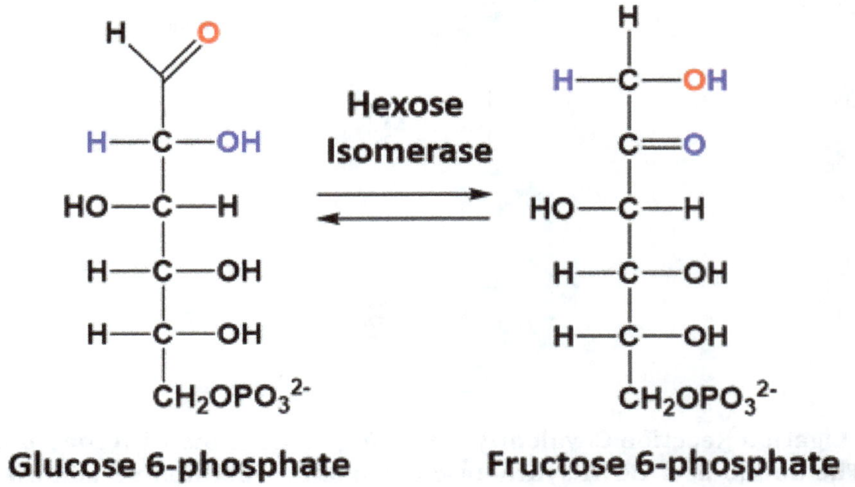

Fig. 9.5 Isomerization of Glucose 6-phosphate to Fructose 6-phosphate.

Ligation Reactions

Ligation reactions use the energy of ATP to join two molecules together. An example of this kind of reaction is the joining of the amino acid with the transfer RNA (tRNA) molecule during protein synthesis. During protein synthesis the tRNA molecules bring each of the amino acids to the ribosome where they can be incorporated into the newly growing protein sequence. To do this, the tRNA molecules must first be attached to the appropriate amino acid. Specific enzymes are available called amino acyl – tRNA synthetases that mediate this reaction.

The synthetase enzymes use the energy of ATP to covalently attach the amino acid to the tRNA molecule. A diagram of this process is shown in Figure 9.6. For each of the 20 amino acids, there is a specific tRNA molecule and a specific synthetase enzyme that will ensure the correct attachment of the correct amino acid with its tRNA molecule.

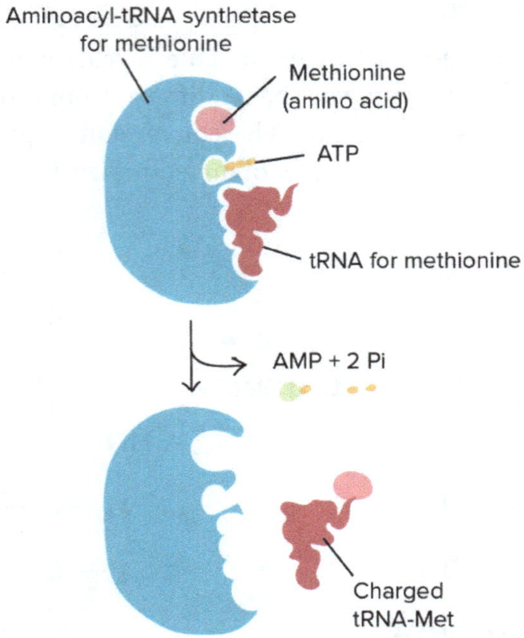

Fig. 9.6 Ligation Reaction Covalently Attaching Methionine with the Appropriate tRNA. The amino-acyl tRNA synthetase enzyme for methionine (shown in blue) covalently attaches methionine (light pink) with the methionine tRNA molecule (dark pink). This reaction requires the energy provided from the breakdown of the ATP molecule into AMP, releasing energy with the breakdown of the phosphate bonds into two inorganic phosphate ions (2 Pi).

Hydrolysis Reactions

The classification of hydrolysis reactions include both the forward reactions that involve the addition of water to a molecule to break it apart or the reverse reaction involving the removal of water to join molecules together, termed dehydration synthesis (or condensation) (Figure 9.7). When water is added to a molecule to break it apart into two molecules this reaction is called hydrolysis.

The term 'lysis' means to break apart, and the term 'hydro' refers to water. Thus, the term hydrolysis means to break apart with water. The reverse of that reaction involves the removal of water from two molecules to join them together into a larger molecule. Since the two molecules are losing water, they are being dehydrated.

Thus, the formation of molecules through the removal of water is known as dehydration synthesis. Since water is also a by-product of these reactions, they are also commonly referred to as condensation reactions. During normal digestion of our food molecules, the major macromolecules are broken down into their building blocks through the process of hydrolysis.

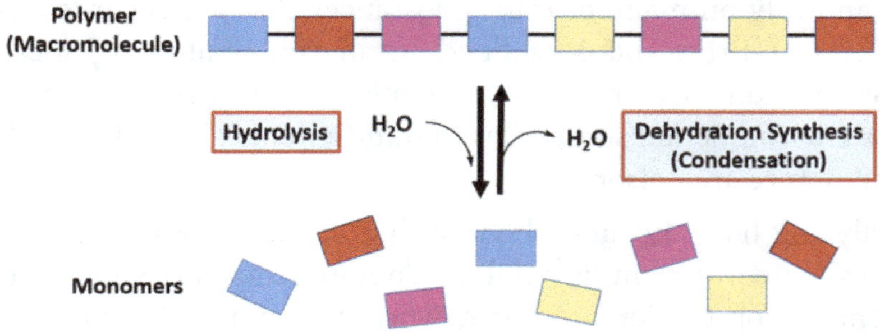

Fig. 9.7 Hydrolysis and Dehydration Synthesis. The reactions of hydrolysis mediate the breakdown of larger polymers into their monomeric building blocks by the addition of water to the molecules. The reverse of the reaction is dehydration synthesis, where water is removed from the monomer building blocks to create the larger polymer structure.

ENVIRONMENTAL CHEMISTRY AND REACTIVITY

Environmental chemistry is an interdisciplinary science that includes atmospheric, aquatic and soil chemistry, as well as heavily relying on analytical chemistry and being related to environmental and other areas of science. Environmental chemistry is the study of chemical processes occurring in the environment which are impacted by humankind's activities.

Definition of Environmental Chemistry

Environmental chemistry describes different chemical processes that happen in air, land, water, and everywhere in the environment and the consequences of human activities upon them. This also includes several parts of chemistry such as "marine chemistry", 'astrochemistry', 'geochemistry', "environmental modeling", "pollution remediation" and "atmospheric chemistry". Environmental chemistry is a scientific study of different biochemical and chemical phenomena that takes place in the environment naturally. The importance of this segment of chemistry is that it examines the risk of several contaminant factors groundwater and soil that harms human health and ecology as well. This study is supposed to expose the reasons for the negative aspects of the environmental harms and the precautions have been suggested based on the study. Human activities are supposed to be reduced for the betterment of the environment and the natural elements are supposed to be utilized for the betterment of nature.

Importance of Environmental Chemistry

Due to its connections to environmental impact, pollutants, environmental management, and the reduction of contamination, environmental chemistry is of high value. Numerous creatures, including humans, are exposed to environmental chemicals and rely on many of these substances that are regarded as nutrients. Volcanoes, forest fires, and lake flooding all contributed significantly to the contamination. Indirectly or directly, organisms have a role in the dispersion of unwelcome environmental substances to some extent. The majority of substances that exist in nature are poisonous.

- Analyzing how chemicals behave in a particular region as well as in the overall environment is vital to comprehend and study the effects of chemicals on the environment. The environment is divided into various sections to address the diversity of natural systems. These are regarded as the components of the actual surroundings that are enclosed by a line dividing them from the outside world.
- The physicochemical characteristics of the chemicals as well as the attributes of environmental compartments describe how they behave in the environment. These two characteristics give insight on the mechanisms underlying chemical and biological processes like hydrolysis, biodegradation, photolysis, and phase transfer processes like sorption and air-water exchange.
- Hydrosphere, sediment, atmosphere, biota, soil, and groundwater are the significant environmental compartments, and it is their significant characteristics and processes that control how chemical pollutants behave. Metal speciation, degradation, and transformation of organic compounds, as well as processes that affect the bioavailability of metals and organic pollutants, are some of the significant environmental processes.
- Examining and addressing environmental challenges is one of the main goals of environmental science. To this end, scientists are examining two key types of relationships between people and the environment. How we use natural resources, such water and plants, is one area of concern.

Fundamentals of environmental chemistry

The consequences of several chemical and natural activities in the environment that change the status of environmental chemistry are major factors of the same. There are four major factors of environmental chemicals such as the geosphere, the hydrosphere, the atmosphere, and the biosphere. All of these factors are the major pillars of the environment that controls the aspects of the environment and control the aspects of environmental chemistry. In this study, a major idea has been focused on based on the consideration of the four major pillars of the environment.

The geosphere, the hydrosphere, the atmosphere, and the biosphere are observed with high efficiency and knowledge of environmental chemistry that detects the changes among these factors. The changes noticed in these four aspects are assessed with great effort to detect the effects of natural and human activities upon nature. The changes that are needed to improve the status of the environmental aspects are supposed to be recommended by the report of environmental chemistry.

Environmental chemistry and ecotoxicology

The survey of different chemical processes that happen in the air, water, land, and everywhere in the environment and the result of human activities upon them is known as environmental chemistry. Environmental chemistry describes different chemical processes that happen in air, land, water, and everywhere in the environment and the consequences of human activities upon them. The toxic nature of the consequences of human activities and other natural aspects deteriorates the wellbeing of the environment that has been revealed through this study.

Ecotoxicology is the mixture of studying "molecular biology", mathematics', analytical chemistry, physiology', toxicology' and ecology'. This compact study exposes the effects of contaminants in nature for the overutilization of pesticides. This study also determines the consequences of the natural and human activities upon the ecosystem, population, and even on the natural communities.

Ecotoxicology helps to clarify the view of uneven human and natural activities and also detects the level of usage of chemicals in nature. This implies the importance of the implication of natural and biopesticides for farming that helps to reduce the negative impacts from the environment.

Atmospheric Pollution

- Tropospheric Pollution:
- Oxides of Sulphur and Nitrogen:

$SO_2 + O_2 \rightarrow 2SO_3$ (Formation of SO_3 from SO_2)

$2NO + O_2 \rightarrow 2NO_2$ (Formation of nitrogen dioxide)

Carbon Monoxide:

$2C + O_2 \rightarrow 2CO$ (Formation of carbon monoxide)

Ozone:

$3O_2 \rightarrow 2O_3$ (In the presence of UV light)

Greenhouse Gases:

$CH_4 + 2O_2 \rightarrow CO_2 + 2H_2O$ (Combustion of methane)

- **Stratospheric Pollution:** Depletion of zone layer due to chlorofluorocarbons (CFCs):

$$CFCl_3 + UV \rightarrow CFCl_2 + Cl$$
$$2Cl + O_3 \rightarrow ClO + O_2$$
$$ClO + O \rightarrow Cl + O_2$$

Water Pollution

- **Hard Water:** Causes of hardness include dissolved calcium and magnesium ions.

$Ca(HCO_3)_2 \rightarrow CaCO_3 + CO_2 + H_2O$ (Calcium bicarbonate forming calcium carbonate)

- **Eutrophication:** Caused by excess nutrients like phosphates and nitrates leading to excessive growth of algae.

$-NO_2-$ + Bacteria $\rightarrow NO_3-$ (Nitrite being converted to nitrate)

Soil Pollution

Pesticides: Example: DDT (Dichlorodiphenyltrichloroethane)

$$C_{14}H_9Cl_5$$

Industrial Wastes

Metals and Non-metals

Example: Hg, Pb, Cd, As and their harmful compounds like HgS, $Pb(CH_3COO)_2$, CdS, As_2O_3

Dyes, Drugs, and Detergents causing water pollution.

Strategies to Control Environmental Pollution

Catalytic Converters

$2CO + 2NO \rightarrow N_2 + 2CO_2$ (Reduction of nitrogen oxide by carbon monoxide)

$2C_8H_{18} + 25O2 \rightarrow 16CO_2 + 18H_2O$ (Combustion of octane)

Wastewater Treatment

Primary Treatment: Physical removal of large impurities.

Secondary Treatment: Biological treatment to remove organic matter.

Tertiary Treatment: Chemical treatment to make the water safe for discharge.

Acid Rain: Acid rain is caused by the release of sulfur dioxide (SO_2) and nitrogen oxides (NO^-) into the atmosphere. These gases react with water vapor and other chemicals in the atmosphere to produce acidic solutions, which then fall as acid rain.

Sulphur Dioxide: Combustion of fossil fuels containing sulphur impurities:

$$S+O_2 \rightarrow SO_2$$

$SO_2 + H_2O \rightarrow H_2SO_3$ (Formation of sulfurous acid)

Nitrogen Oxides: Due to vehicular exhaust and industrial processes:

$$N_2 + O_2 \rightarrow 2NO_2$$

$$2NO + O_2 \rightarrow 2NO_2$$

$NO_2 + H2O \rightarrow HNO_3 + NO$ (Formation of nitric acid)

Effects: $CaCO_3 + H2SO_4 \rightarrow CaSO_4 + H_2O + CO_2$ (Limestone reacting with sulfuric acid)

INDUSTRIAL PROCESSES AND REACTIVITY

Types of Industrial Processes and Materials Used

While not an exhaustive list, the following are the more common types of manufacturing processes.

Chemical Processes

Several chemical processes create the necessary materials in manufacturing. These include cement, steel, aluminum and fertilizer. The chemical process for cement is called calcination. For steel, the process is called smelting, which is done inside a blast furnace. The Hall Heroult process is used to make aluminum, while the Haboer process produces fertilizer. There are also chemical processes to combine materials called pyroprocessing. Disinfection is a chemical treatment that kills bacteria and viruses.

Electrolysis

The use of electricity is another industrial process that affects materials and is used for plating and separating metals. These include gilding, electroplating, anodization, electrowinning, electropolishing, electrofocusing, electrolytic process, electrophoretic deposition, electrotyping and metalizing, platting and spin control.

Cutting

Cutting is an industrial process that is used in shearing, sawing, plasma cutting, water-jet cutting, oxyacetylene cutting, electrical discharge machining (EDM) and laser cutting. Machining also falls into this category type. It refers to the mechanical cutting and shaping of metal, which results in the loss of some material.

Molding

Here, materials are made by liquifying them and shaping them in molds as they cool and become solid. This is done in casting, sand casting, sintering, powder metallurgy, blow molding and compression molding.

Separation (Refining of Materials)

This process takes impure materials and purifies them by separation, which creates a product that can be used in production. This is done by reducing the size of particles by comminution, separating materials with froth flotation and the flotation process. There is also liquid-liquid extraction, which dissolves one substance in the other, and the Frasch process, which extracts molten sulfur from the ground.

There are, of course, many other examples of industrial processes from distillation, which purifies volatile substances through evaporation and condensation and metalworking. Metalworking, also called fabrication, includes processes such as smelting, forging, casting, steelmaking, progressive stamping, stamping, hydroforming, sandblasting, soldering, brazing, welding, tumble polishing, precipitation hardening, work hardening, case hardening, die-cutting and using an electric arc furnace.

Industrial Process Examples

Depending on the business, one kind of industrial process can be more effective than another. However, there are many different industrial process examples. The most common types of manufacturing processes include casting, molding, forming, joining, plating and additive. These can be found in the examples below.

Repetitive Manufacturing

This type of industrial manufacturing process is used when repeating the production of a product. It features production lines that are working on similar products or components all year round. This means there is little changeover or setup required, though production will reflect client demand. Companies that use this industrial process example include those that produce electronic goods, automotive and durable consumer goods, etc.

Job Shop Manufacturing

Here, parts are used in small-batch, custom products. These are made-to-order items for clients with workstations that are usually dedicated to one specific product. Think of a custom shoemaker or commercial printing press, which can change a product to match the client's request. Some machine shops use this type of industrial process and make parts for the aviation industry. However, it also presents software development, helping with workflow. Some businesses move from job shop to repetitive manufacturing when they need to scale for greater productivity.

Discrete Manufacturing

This uses an assembly line, but with a greater variety than the repetitive manufacturing example. Therefore, it can have more recurrent changeover and variation in the process. This suits a manufacturer with many styles, sizes and modifications of their product. The downside is that production usually takes longer due to more setup and removal. This type of industrial process can be found in automobile and aircraft manufacturers, as well as those who create clothing, medical devices, toys and smartphones.

Batch Process Manufacturing

Driven by customer demand and the availability of raw materials, this type of manufacturing is similar to discrete and job shop processes. Usually, this means that one batch is enough for client needs and when completed, the equipment is cleaned up and the company waits until a new need for product arises at which time they'll start up production again. Industries that use this type of industrial manufacturing process include food production, newspaper printing, bookbinding and pharmaceuticals.

Continuous Process Manufacturing

As with repetitive manufacturing, continuous process manufacturing happens all the time. The two types of industrial manufacturing processes differ in that the raw materials are usually gases, powders, liquids or slurry. This example can be found in industries such as oil refining, metal smelting, paper production and food products that are similar to tomato sauce, juice and peanut butter.

3-D Printing

3-D printing uses several composites and materials, such as plastics and metals, to create three-dimensional goods layer by layer. This is done through digital means rather than physical labor or mechanization. Benefits to this type of manufacturing include reduced financial capital, raw materials and waste, though the process is expensive. It can also be used to create prototypes before committing to larger-scale projects and is used to make medical and dental devices, prosthetic limbs, firearms, shows, musical instruments and even buildings.

Kanban Board Template

This free kanban board template can help manufacturers monitor each stage of their production process. To do so, you can use kanban board columns to represent the industrial processes that are executed to manufacture a product such as cutting, molding and painting. Then you can use kanban cards to visualize production orders and then move them throughout the production stages as each step is completed.

Industrial processes are procedures involving chemical, physical, electrical, or mechanical steps to aid in the manufacturing of an item or items, usually carried out on a very large scale. Industrial processes are the key components of heavy industry.

Chemical processes by main basic material

Certain chemical process yield important basic materials for society, e.g., (cement, steel, aluminum, and fertilizer). However, these chemical reactions contribute to climate change by emitting carbon dioxide, a greenhouse gas, through chemical reactions, as well as through the combustion of fossil fuels to generate the high temperatures needed to reach the activation energies of the chemical reactions.

Cement (the paste within concrete)

- **Calcination** – Limestone, which is largely composed of fossilized calcium carbonate ($CaCO_3$), breaks down at high temperatures into useable calcium oxide (CaO) and carbon dioxide gas (CO_2), which gets released as a by-product. This chemical reaction, called calcination, figures most prominently in creating cement (the paste within concrete). The reaction is also important in providing calcium oxide to act as a chemical flux (removal of impurities) within a blast furnace.

$$CaCO_3(s) \rightarrow CaO(s) + CO_2(g)$$

Steel

- **Smelting** – Inside a blast furnace, carbon monoxide (CO) is released by combusting coke (a high-carbon derivative of coal) and removes the undesired oxygen (O) within ores. CO2 is released as a by-product, carrying away the oxygen and leaving behind the desired pure metal. Most prominently, iron smelting is how steel (largely iron with small amounts of carbon) is created from mined iron ore and coal.

$$Fe_2O_3(s) + 3\,CO(g) \rightarrow 2\,Fe(s) + 3\,CO_2(g)$$ [1]

Aluminium

- **Hall–Héroult process** – Aluminium oxide (Al_2O_3) is smelted with coke (C) in a high-temperature electrolysis reaction, yielding the desired pure aluminium (Al) and a mixture of CO and CO_2.

$$Al_2O_3(s) + 3\ C(s) \rightarrow 2\ Al(s) + 3\ CO(g)$$
$$2\ Al_2O_3(s) + 3\ C(s) \rightarrow 4\ Al(s) + 3\ CO_2(g)$$

Fertilizer

- **Haber process** – Atmospheric nitrogen (N_2) is separated, yielding ammonia (NH_3), which is used to make all synthetic fertilizer. The Haber process uses a fossil carbon source, generally natural gas, to provide the CO for the water–gas shift reaction, yielding hydrogen (H_2) and releasing CO_2. The H_2 is used to break the strong triple bond in N_2, yielding industrial ammonia.

$$CH_4(g) + H_2O(g) \rightarrow CO(g) + 3\ H_2(g)$$
$$CO(g) + H_2O(g) \rightarrow H_2(g) + CO_2(g)$$
$$N_2(g) + 3\ H_2(g) \rightarrow 2\ NH_3(g)$$

Other chemical processes

- Disinfection – chemical treatment to kill bacteria and viruses
- Pyroprocessing – using heat to chemically combine materials, such as in cement

10

ADVANCES IN UNDERSTANDING CHEMICAL REACTIVITY

A chemical reaction is a process in which reactants react chemically and convert into products by chemical transformation. For example, Respiration – we inhale oxygen which reacts with glucose and produces carbon dioxide, water and energy. The reaction is given below.

$$C_6H_{12}O_6 + 6O_2 \rightarrow 6CO_2 + 6H_2O + \text{Energy}$$

Characteristics of a Chemical Reaction

Chemical reactions show one or more than one of the following characteristics.

- Evolution of gas
- Change in colour
- Change in temperature
- Change in energy
- Formation of precipitate
- Change in state

Now understand these characteristics in detail.

- Evolution of Gas – Some chemical reactions take place with the evolution of gas. For example, when zinc reacts with hydrochloric acid, hydrogen gas is evolved with the formation of zinc chloride.

Reaction - $Zn + 2HCl \rightarrow ZnCl_2 + H_2$

- **Change in Colour** – Some chemical reactions are accompanied by a change in colour. For example, when colourless lead nitrate reacts with potassium iodide, it forms a yellow precipitate of lead iodide and colourless potassium nitrate.

Reaction – $Pb(NO_3)_2 + 2KI \rightarrow PbI_2 + 2KNO_3$

Colourless Colourless Yellow ppt colourless

- **Change in Temperature** – Some chemical reactions are accompanied by a change in temperature. For example, some reactions are exothermic in nature, while some are endothermic.
- **Change in Energy** – Chemical reactions often involve a change in energy. During a chemical reaction, energy can be evolved or be absorbed. For example, in combustion reactions, energy is released, such as the burning of a fuel.
- **Formation of Precipitate** – Some chemical reactions are accompanied by the formation of the precipitate. For example, barium chloride reacts with sodium sulphate and forms sodium chloride and precipitate of barium sulfate.

Reaction – $BaCl_2 + Na2SO_4 \rightarrow BaSO_4 + NaCl$

Precipitate

- **Change in State** – Some chemical reactions are accompanied by a change in state. For example, ammonia gas reacts with hydrogen chloride gas and forms solid ammonium chloride crystals.

Reaction – $NH_{3(g)} + HCl_{(g)} \rightarrow NH_4Cl_{(s)}$

Types of Chemical Reactions

Chemical reactions are of many types. Here, we are covering the most common eight types of chemical reactions, which are part of the class X chemistry syllabus as well.

- Decomposition reaction
- Combination reaction
- Combustion reaction
- Neutralization reaction
- Single displacement reaction
- Double displacement reaction
- Precipitation reaction
- Redox reaction
- **Decomposition Reaction** – In a decomposition reaction, molecules or compounds break down into two or more than two simpler chemically new substances. For example, electrolysis of water. In the electrolysis of water,

water breaks down into hydrogen and oxygen, which show completely different properties than water.

Reaction - $2H_2O$ electricity → $2H_2 + O_2$

- **Combination Reaction** – In a combination reaction, two or more molecules are combined together chemically to form a new substance (compound). Combination and decomposition reactions are opposite of each other. For example, when we burn magnesium ribbon (or magnesium), it gives grey-black ash of magnesium oxide.

Reaction – $Mg + O_2 \rightarrow MgO$

- **Combustion Reaction** – It is an exothermic reaction that releases energy, generally in the form of heat. It is a reaction between fuel and an oxidant (generally atmospheric oxygen) that produces smoke, water and heat generally. For example, when we burn methane, it gives carbon dioxide and water.

Reaction – $CH_4 + 2O_2 \rightarrow CO_2 + 2H_2O$

- **Neutralization Reaction** – In these reactions, acid and base react with each other and form salt and water. For example, hydrochloric acid reacts with sodium hydroxide (base) and forms sodium chloride (salt) and water.
- **Single Displacement Reaction** – In these reactions, more reactive metal displaces less reactive metal from its salt. In these reactions, products can be determined through reactivity series. Reactivity series is a series in which elements are arranged in decreasing order of their reactivity. It means the elements present at the top of this reactivity series are more reactive than the elements present at the bottom.

The reaction of potassium with magnesium chloride is an example of a single displacement reaction. In this reaction, potassium displaces magnesium from its salt because potassium is more reactive than magnesium. Potassium is present at the top of the reactivity series and is the most reactive element.

Reaction – $2K + MgCl_2 \rightarrow 2KCl + Mg$

- Double displacement reaction – In these reactions, two aqueous ionic compounds exchange their ions (mostly cations) and produce two new compounds. For example, potassium nitrate reacts with aluminium chloride and forms aluminium nitrate and potassium chloride.

Reaction – $KNO_3 + AlCl_3 \leftrightarrow Al(NO_3)_3 + KCl$

- Precipitation Reaction – In these reactions, an insoluble precipitate is formed. In precipitation reactions, two soluble salts in aqueous solutions are combined and form an insoluble precipitate.

Reaction – $AgNO_{3(aq)} + KCl_{(aq)} \leftrightarrow AgCl + KNO_{3(aq)}$

- **Redox Reaction** – Those chemical reactions in which oxidation and reduction take place simultaneously are called redox reactions. Oxidation is the addition of oxygen, while reduction is the addition of hydrogen (or removal of oxygen). We have already discussed oxidation and reduction in a separate article with the title 'Oxidation and Reduction'. You can refer to this article for a detailed explanation of oxidation and reduction.

The reaction of copper oxide with hydrogen is an example of a redox reaction. In this reaction, hydrogen has undergone oxidation by gaining oxygen atoms while copper oxide has undergone reduction by removing oxygen.

List of Common Chemical Reactions

S.No.	Chemical reaction	Equation
1.	Electrolysis of water	$2H_2O$ electricity→ $2H_2 + O_2$
2.	Rusting of iron	$4Fe + 3O_2 \rightarrow 2Fe_2O_3$
3.	The reaction of quicklime (CaO) with water	$CaO + H_2O \rightarrow Ca(OH)_2 Ca(OH)_2$ is known as slaked lime.
4.	Photosynthesis	$6CO_2 + 6H_2O \rightarrow C_6H_{12}O_6 + 6O_2$ (reaction takes place in the presence of sunlight and chlorophyll)
5.	Respiration	$C_6H_{12}O_6 + 6O_2 \rightarrow 6CO_2 + 6H_2O$
6.	Combustion of hydrogen – Reaction of hydrogen gas with pure hydrogen. This is an exothermic reaction.	$2H_2 + O_2 \rightarrow 2H_2O$ + energy
7.	Decomposition of $FeSO_4$	$2FeSO_4$ +heat ® $Fe_2O_{3(s)} + SO_{2(g)} + SO_{3(g)}$($Fe_2O_{3(s)}$ is green in color)
8.	Decomposition of lead nitrate – In this reaction, brown fumes of nitrogen dioxide gas are formed with the yellow coloured residue of lead oxide.	$2Pb(NO_3)_2 \rightarrow 2PbO + 4NO_2 + O_2$
9.	Displacement reaction of iron and copper sulfate – Iron is more reactive than copper, so it displaces copper from its salt and produces iron surfate.	$Fe + CuSO_4 \rightarrow FeSO_4 + Cu$

CUTTING-EDGETECHNIQUESINSTUDYINGREACTIONMECHANISMS

One key aspect of innovation in chemical process research is the integration of cutting edge technology. Advanced tools such as high-throughput experimentation, machine learning algorithms, and computational modelling have revolutionised the way chemists approach problem-solving. These technologies enable researchers to analyse vast amounts of data, predict chemical behaviours, and accelerate the discovery of new processes.

Collaboration plays a vital role in fostering innovation within process research. Interdisciplinary partnerships between chemists, engineers, and material scientists facilitate the exchange of ideas and expertise, leading to synergistic solutions that address complex challenges. By working together, researchers can leverage diverse perspectives and skills to overcome obstacles and drive progress in the field.

Strategies for Optimising Chemical Processes

Optimising chemical processes is a crucial aspect of chemical development. It involves enhancing reaction yields, increasing selectivity, and minimising undesired by-products.

One strategy for optimisation is the use of advanced analytical techniques. By employing sophisticated tools such as spectroscopy and chromatography, scientists can monitor reactions in real-time, allowing for precise control and adjustment.

One fascinating aspect of employing advanced analytical techniques is the ability to delve deep into the molecular world of chemical reactions. Spectroscopy, for example, can provide valuable insights into the electronic structure of molecules involved in the process, aiding in understanding reaction mechanisms and identifying potential bottlenecks. Similarly, chromatography not only helps in separating and analysing reaction components but also plays a crucial role in purifying products to meet industry standards.

Another approach is the development of intelligent reactor designs. The integration of automation, sensors, and feedback systems enables researchers to fine-tune reaction conditions, maximizing efficiency and reducing waste.

When it comes to intelligent reactor designs, the use of computational fluid dynamics (CFD) has revolutionised the way researchers approach chemical process optimisation. By simulating fluid flow, heat transfer, and chemical reactions within a reactor, scientists can predict and optimise mixing patterns, residence times, and temperature gradients. This level of detailed analysis allows for the creation of bespoke reactor designs tailored to specific chemical processes, ultimately leading to improved performance and productivity.

The incorporation of green chemistry principles, such as the use of renewable resources and the minimization of hazardous materials, is essential for achieving sustainable and environmentally friendly chemical processes.

Green chemistry principles not only focus on the end products of chemical process development but also emphasise the importance of every step in the production chain. From the sourcing of raw materials to the disposal of waste, each stage is carefully evaluated to minimise environmental impact. By adopting these principles, chemical engineers can develop processes that not only meet performance targets but also contribute to a greener and more sustainable future for the industry and the planet as a whole.

Importance of Safety Protocols in Chemical Development

In the realm of chemical development, safety protocols are of utmost importance. Chemical reactions can be hazardous, involving highly reactive substances and volatile conditions. Thus, stringent safety measures must be implemented to protect researchers and the surrounding environment.

Proper training, lab equipment maintenance, and adherence to safety guidelines are essential to minimise risks. Additionally, the development of safety assessment methods and the continuous improvement of safety protocols are vital for ensuring the well-being of those involved in chemical development.

By prioritising safety, researchers can carry out experiments with confidence, allowing for the exploration of new frontiers in chemical reaction mechanisms.

It is crucial for chemical development facilities to have designated safety officers who oversee the implementation of safety protocols and conduct regular safety audits. These officers play a key role in identifying potential hazards, providing safety training to staff, and ensuring that emergency response plans are in place.

The use of personal protective equipment (PPE) such as lab coats, gloves, and safety goggles is non-negotiable in any chemical laboratory setting. PPE serves as a physical barrier between researchers and hazardous substances, reducing the risk of exposure and injury.

Investing in state-of-the-art safety technology, such as fume hoods, emergency showers, and eyewash stations, is essential for creating a safe working environment in chemical development laboratories. These safety measures act as a first line of defence in the event of accidents or chemical spills, providing researchers with the necessary tools to mitigate risks effectively.

Trends in Green Chemistry for Sustainable Processes

In recent years, there has been a growing emphasis on sustainability in chemical development. Green chemistry, with its focus on minimising environmental impact and promoting sustainable practices, is a prominent trend in the field.

Researchers are actively seeking greener alternatives to traditional processes, opting for renewable feedstocks, developing catalytic systems with reduced toxicity, and implementing energy-efficient strategies.

Not only does green chemistry improve process efficiency and cut costs, but it also helps the environment.. By embracing sustainable practices, chemical development can pave the way for a brighter, cleaner future.

The concept of atom economy plays a crucial role in green chemistry. Atom economy focuses on maximising the incorporation of all starting materials into the final product, thereby reducing waste generation. This approach not only minimises the environmental impact but also improves the overall efficiency of chemical reactions, leading to more sustainable processes.

Leveraging Technology in Process Research

Technology plays a pivotal role in advancing process research in the realm of chemical development. Cutting-edge instruments, automation, and computational tools empower scientists to explore, analyse, and optimise reactions with unprecedented precision.

High-throughput screening techniques allow for the rapid screening of thousands of reaction conditions, accelerating the discovery of new processes. Computational modelling provides insights into reaction mechanisms, aiding in the design of more efficient catalysts and reaction pathways.

Data analytics and machine learning algorithms enable researchers to extract valuable information from vast datasets, facilitating the identification of key factors influencing reaction rates.

By embracing technological advancements, process research becomes a dynamic and transformative field, propelling the understanding of chemical reaction mechanisms to new heights.

One fascinating aspect of the integration of technology in process research is the use of flow chemistry systems. These systems enable continuous processing of reactions, offering improved control over reaction parameters and enhanced safety compared to traditional batch reactions. Flow chemistry also allows for rapid optimization of reaction conditions and the synthesis of complex molecules in a more streamlined and efficient manner.

The development of in-situ monitoring techniques, such as real-time spectroscopic analysis and online mass spectrometry, has revolutionised the way researchers observe and understand chemical reactions. These monitoring tools provide instant feedback on reaction progress, allowing for real-time adjustments and enhancing the overall efficiency of process development.

Regulatory Considerations in Chemical Development

Chemical development is not only driven by scientific curiosity but also guided by regulatory frameworks. Governments and organisations worldwide have established stringent regulations to ensure the safety and environmental sustainability of chemical processes.

Researchers must navigate these regulations to ensure compliance while pushing the boundaries of innovation. By collaborating with regulatory bodies, scientists can explore mutually beneficial ways to enhance safety, sustainability, and the overall impact of chemical development.

Regulatory considerations are integral to fostering responsible and ethical research practices, ensuring that the benefits of chemical reactions are harnessed without compromising public health or the environment.

One key aspect of regulatory considerations in chemical development is the need for thorough risk assessments. Before a new chemical process can be implemented on a larger scale, researchers must conduct comprehensive risk assessments to identify potential hazards and develop strategies to mitigate them. This proactive approach not only safeguards the well-being of workers and the surrounding community but also helps in preventing environmental contamination.

Regulatory frameworks often require detailed documentation and reporting throughout the chemical development process. This documentation serves as a crucial tool for monitoring and evaluating the impact of chemical processes on human health and the environment. By maintaining accurate records and transparent reporting practices, researchers can demonstrate their commitment to compliance and accountability, ultimately contributing to the advancement of sustainable and responsible chemical development.

Collaborative Approaches in Advancing Chemical Processes

No field exists in isolation, and chemical development is no exception. Collaborative approaches, involving partnerships between academia, industry, and government entities, play a vital role in advancing chemical processes.

Collaboration within the realm of chemical development is a multifaceted endeavour, bringing together diverse expertise and resources to tackle complex challenges. Academic institutions, with their focus on research and innovation, provide the foundational knowledge and scientific curiosity that underpins advancements in the field. Industry partners, on the other hand, offer practical insights into real-world applications, ensuring that research findings can be translated into tangible solutions that benefit society.

Government entities play a crucial role in shaping the landscape of chemical development through funding opportunities and regulatory frameworks. By supporting collaborative initiatives, governments can catalyse innovation, address pressing environmental and societal issues, and drive economic growth through the development of new technologies and processes.

Through collaboration, researchers gain access to valuable resources, expertise, and funding. Industry partners provide real-world applications and scale-up capabilities, while academic institutions contribute fundamental research and innovation.

Government support, through grants and policies, spurs research and ensures that chemical development aligns with societal needs and priorities.

By fostering collaboration, chemical development transcends individual efforts, leading to transformative advancements that benefit society as a whole.

Factors Influencing Reaction Rates

Understanding the factors that influence reaction rates is essential for controlling and optimizing chemical processes. Several key factors must be considered:

- **Temperature:** Increasing temperature generally increases reaction rates by providing energy to overcome activation barriers.
- **Concentration:** Higher concentrations of reactants increase the rate of collisions, leading to more frequent successful reactions.
- An alternative reaction pathway with a lower activation energy can be provided by catalysts, which expedite processes.
- **Surface area:** Increasing the surface area of reactants, such as through particle size reduction, enhances contact between molecules, leading to faster reactions.
- **Pressure:** In certain reactions, higher pressures can increase reaction rates by favoring the formation of desired products.

By carefully manipulating these factors, researchers can gain precise control over reaction rates, enabling the development of efficient and cost-effective chemical processes.

Common Reaction Mechanisms in Organic Chemistry

Organic chemistry, a branch of chemistry dedicated to the study of carbon-based molecules, is rich with fascinating reaction mechanisms. Let us explore some of the most common mechanisms:

- **Nucleophilic Substitution:** In this reaction, a nucleophile replaces a leaving group, resulting in the substitution of one atom or group with another.
- **Electrophilic Addition:** Electrophiles react with nucleophiles, resulting in the addition of atoms or groups to an unsaturated molecule.
- **Elimination:** In elimination reactions, functional groups are removed from a molecule, typically resulting in the formation of a double bond.

- **Radical Reactions:** Radical reactions involve the formation and reaction of highly reactive species known as radicals, often initiated by the breaking of covalent bonds through the input of energy, such as heat or light.
- **Reduction and Oxidation:** These reactions involve the gain or loss of electrons, resulting in the alteration of oxidation states of atoms.

By understanding and manipulating these common reaction mechanisms, chemists can unlock a vast array of possibilities to create new compounds and develop innovative chemical processes.

Techniques for Studying Reaction Mechanisms

Studying reaction mechanisms is a complex task that requires the use of various techniques and approaches. Some of the most commonly employed techniques include:

- **Spectroscopy:** Spectroscopic methods, including NMR and infrared spectroscopy, provide invaluable information about molecular structures and vibrational modes.
- **Mass Spectrometry:** Mass spectrometry enables the identification and characterization of molecules by separating and analysing ionised fragments.
- **Reaction Kinetics:** Measurement of reaction rates and the study of reaction kinetics shed light on the sequence of steps involved in a chemical reaction.
- **Computational Modelling:** Computational methods, such as density functional theory (DFT) and molecular dynamics simulations, allow for the simulation and prediction of reaction mechanisms.

Chemical transformations can be better understood with the use of these methods, which combine experimental data with theoretical models to reveal the intricate workings of reaction mechanisms.

Future Trends in Chemical Reaction Mechanisms

The field of chemical reaction mechanisms is ever-evolving, driven by scientific curiosity and the need for innovative solutions. Several future trends are expected to shape the landscape of chemical development:

- **Advanced Computational Tools:** The development of increasingly powerful and accessible computational tools will revolutionise the study of reaction mechanisms, enabling researchers to explore complex systems with unprecedented accuracy.
- **Sustainable Processes:** The integration of green chemistry principles will continue to be a focal point, driving the development of sustainable

processes that minimise environmental impact and promote a circular economy.

- **Artificial Intelligence:** The use of artificial intelligence and machine learning algorithms will aid in the prediction and design of reactions, accelerating the discovery of new catalysts and reaction pathways.
- **Multi-disciplinary Collaboration:** The convergence of various scientific disciplines, such as chemistry, biology, and materials science, will lead to synergistic advancements in chemical reaction mechanisms.

As chemistry continues to progress, we can expect exciting discoveries and transformative breakthroughs in the realm of chemical reaction mechanisms.

Steps Involved in Reaction Mechanisms

Reaction mechanisms are intricate sequences of steps that detail the progress of a chemical reaction. Although each reaction is unique, common steps often appear:

- **Initiation:** In this step, a reactive intermediate or species is formed, often through the input of energy, which initiates the reaction.
- **Propagation:** The reactive species generated in the initiation step reacts with other molecules to form new products and additional reactive species that propagate the reaction.
- **Termination:** Eventually, the reactive species responsible for propagating the reaction are consumed, leading to the termination of the reaction. Multiple termination steps may occur, resulting in the formation of various products.

By dissecting and understanding these steps, scientists can gain insights into the underlying mechanisms of chemical reactions, enabling them to design more efficient and selective processes.

Role of Catalysts in Chemical Reactions

Chemical reactions rely on catalysts, which improve the efficiency of reactant-product transformations by changing reaction speeds.

A catalyst works by providing an alternative reaction pathway that involves lower activation energy. It allows reactions to proceed under milder conditions, reducing energy requirements and facilitating selectivity.

Catalysts can be homogeneous, where the catalyst and reactants are in the same phase, or heterogeneous, where the catalyst is present in a different phase. Additionally, enzymes, biological catalysts, play a vital role in numerous biochemical reactions. By harnessing the power of catalysts, scientists can unlock new reaction possibilities and develop processes that are more energy-efficient and economically viable.

Applications of Understanding Reaction Mechanisms

The development of new materials, pharmaceuticals, and sustainable processes are all impacted by our growing knowledge of reaction mechanisms, which has far-reaching implications across many industries.

By comprehending reaction mechanisms, chemists can design novel materials with tailored properties, such as conducting polymers or catalysts with enhanced catalytic activity.

In the field of pharmaceuticals, understanding reaction mechanisms is crucial for the development of new drugs and the optimization of synthetic routes. It allows for the design of efficient processes that yield high-quality medications.

In sustainable chemistry, understanding reaction mechanisms is key to developing eco-friendly processes that minimise waste, energy consumption, and environmental impact.

The applications of understanding reaction mechanisms extend far beyond these examples, permeating every aspect of chemical development, shaping the world we live in and propelling us towards a brighter future.

COMPUTATIONAL APPROACHES TO PREDICTING REACTIVITY

The discovery of chemical reactions is a primary research area that can provide new synthetic methods to create molecules in organic chemistry. Based on the developed methods, synthetic plans can be constructed for the preparation of complex molecules such as drug candidates or functional materials. However, the development of synthetic methodologies usually requires a number of experiments, mainly because the process relies on trial and error as well as on the intuition of organic chemists, which generates unwanted waste and inflates costs.

Moreover, unexpected reactions are also often found during the research campaign, which may change the direction of the research project and are recognized as serendipitous findings that depend both on the researchers' intuition and on luck. Therefore, an efficient and systematic process without unnecessary experiments has long been desired, in line with the recent trends of research automation and digitalization.

For this purpose, extensive investigations into the application of experimental or informatic technologies for methodology development have been conducted. For example, high-throughput experimentation (HTE), a technique widely used in medicinal chemistry for drug discovery that performs multiple experiments in parallel and thus accelerates the screening process, has been applied to investigate chemical reactions. This technique enables the combinatorial screening of reaction parameters, including reagents, catalysts, or additives, leading to conditions that provide better yields or selectivities in known chemical reactions.

Although HTE has been proven to be effective in discovering certain unknown chemical reactions, it still requires a large number of experiments (typically >1000). For the systematic development of chemical reactions, data-driven approaches have also made significant progress with the advancement of chemoinformatic techniques, in which the experimental results are correlated to reaction parameters in order to create models that capture data trends, understanding of mechanistic insights, and prediction of reaction outcomes.

This approach has evolved from the Brønsted catalysis law, a pioneering study in organic chemistry that established a quantitative relationship between acidity and reaction rate, to the data-driven optimization of reaction parameters that include not only conventionally effective factors such as the solvent and temperature, but also the substrate or catalyst structures, to improve the yield or selectivity.

Despite the significant progress in modeling chemical reactions, applications for discovering unknown reactions still remain underdeveloped, and the models developed in reported studies are currently limited to predicting the optimal conditions for known reactions.

Quantum chemical calculations have also been used to investigate the properties of molecules and reaction mechanisms. Since the 1960s, significant advances in computational chemistry, such as the launch of the Gaussian program and the development of density functional theory (DFT), have made the use of quantum chemical calculations more common in organic chemistry. The technology and its accuracy continue to improve, enabling calculations that reveal detailed reaction processes, ranging from entire catalytic cycles to biological reaction pathways involving enzymatic catalysis.

Consequently, quantum chemical calculations are at present primarily employed to gain mechanistic insight into known chemical reactions. Due to the ability of quantum chemical calculations to be performed on even unknown reaction processes, their application in the prediction of reactions holds significant potential for methodology development. In biochemistry, analogous studies are often conducted to simulate protein- protein interactions or docking simulations between small molecules and target protein binding pockets using computational methods such as molecular dynamics, molecular mechanics, or their hybrid methods.

Unlike such simulations, which primarily focus on a single chemical process, the prediction of chemical reactions requires consideration of potential competitive reaction pathways; this encompasses the calculation of the relative energies of potential transition states and the resulting products using quantum-chemical calculations. It should be noted some such processes, e.g., those involving dynamical bifurcations, cannot be predicted accurately based solely on these parameters. This perspective highlights the development of synthetic methodologies based on computational predictions using quantum chemical calculations (Fig. 10.1).

While other reviews on this topic have mainly focused on improving yields or selectivities based on existing results, this perspective places greater emphasis on research in which prediction has significantly contributed to the discovery of new methodologies, including unexplored selectivity, catalyst design, and new synthetic routes toward complex molecules.

We also compare the current calculation methods used to explore chemical reaction pathways, which are critical for predicting reactions, and discuss the potential of modern methods to guide computation-based methodology development. Although such predictive strategies have not yet reached a mature level of practical use, they promise substantial potential to evolve into a next-generation tool. Finally, this perspective discusses the hurdles that remain to be overcome, along with an outlook for the future.

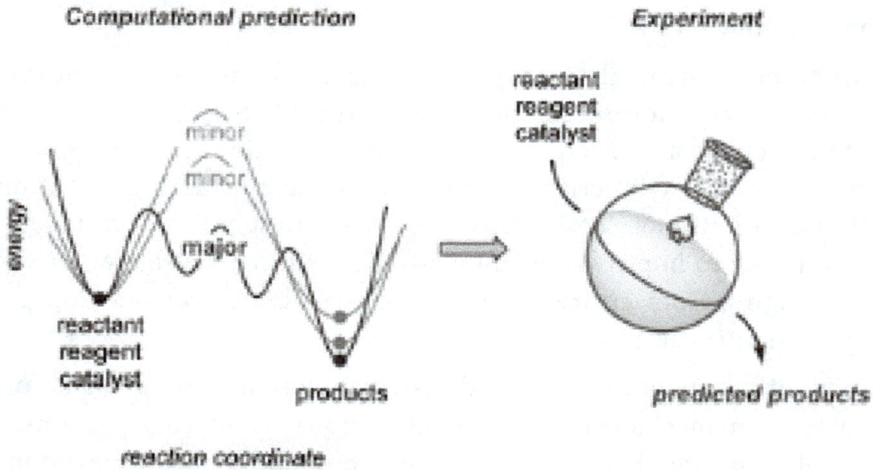

Fig. 10.1 Conceptual illustration of the computation-based development of reactions.

Reaction development using computational predictions of researcher-guided reaction pathways

Pericyclic reactions

In 1987, the predictive capability of quantum chemical calculations based on transition states was first reported by Houk and co-workers in the field of synthetic organic chemistry with the successful prediction and subsequent experimental validation of the conrotatory electrocyclic ring-opening of 3-formylcyclobutene (Fig. 10.2). During their study on the ring-opening of substituted cyclobutenes, the authors found that the reaction of C3- substituted cyclobutenes such as 3-methyl-, 3-chloro-, 3-acetoxy-, and 3-ethoxycyclobutene exhibited a preference for the outward rotation of the substituent, resulting in the selective formation of the E-alkene.

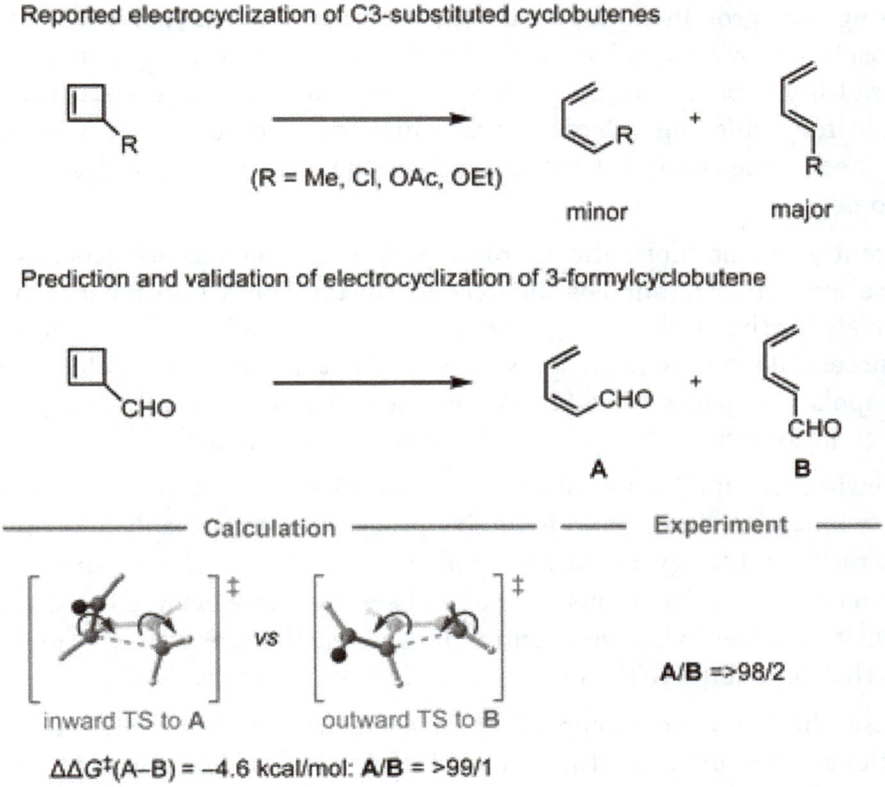

Fig. 10.2 Computational prediction of the ring-opening of 3-formylcyclobutene; calculations were performed at the HF/6-31G(d)//HF/3-21G level.

The authors also investigated the analogous reaction of 3-formylcyclobutene using quantum chemical calculations, which indicated that this substrate can be expected to selectively undergo inward rotation (DDG‡ = -4.6 kcal mol-1) due to the participation of the p*c =c orbital in the stabilization of the HOMO sC–C orbital of the inward transition state, which stands in contrast to the selectivity of the aforementioned example. This counterintuitive prediction was subsequently validated experimentally. These findings prompted further investigations to understand the reaction mode. This study demonstrated that computations could provide predictions that are contrary to human intuition.

Computational design of organocatalysts

The computational design of catalysts is the holy grail in organic chemistry, as conventional catalyst development traditionally relies on a trial-and-error approach based on empiricism. This approach often requires a large number of experiments including the preparation of catalyst candidates and the evaluation of their catalytic activity. To transition from an intuitive approach to a systematic design, studies using computational methods have emerged as one of the most promising and prominent topics in this research field. In particular, catalytic enantioselective reactions have attracted the attention of many researchers due to their utility in producing optically active molecules and their relatively simple principle for achieving selectivity, in which enantiomeric excess is typically determined by the energy difference between the transition states leading to each enantiomer.

Recently, chemoinformatic approaches, in which models are generated using the experimental results and parameters of each target reaction, have been studied extensively for the prediction of the performance of chiral catalysts. Some studies have successfully developed models capable of predicting enantioselectivity, even in extrapolation spaces, allowing for the identification of catalysts that provide higher enantiomeric excess compared to others in the model.

Likewise, the application of quantum chemical calculations in the design of catalysts for controlling enantioselective processes is undeniably attractive, as it offers a rational strategy for the development of catalyst scaffolds. Conventionally, quantum chemical calculations have often been used to elucidate the structure of the transition states in known enantioselective reactions, as well as to identify the factors that contribute to the induction of the observed selectivities.

These abilities have prompted chemists to use them for enantioselectivity predictions, whereby most studies have hitherto focused on the modification of substituents on catalysts to improve known selectivities. In contrast, Houk, Tanaka, Barbas, and co-workers have jointly reported a remarkable design achievement in the field of enantioselective organocatalysis (Fig. 10.3).

Typically, proline-catalyzed Mannich reactions provide an enantioenriched product with syn-selectivity, in which the catalyst dictates the conformation of the chair-like transition state, assisted by the hydrogen-bonding interaction between the imine substrate and the carboxylic acid moiety of proline. In their computational study, a catalyst bearing a methyl group at the C2 position and a carboxylic acid at the C4 position on pyrrolidine was expected to favor the transition state leading to the formation of the anti-product, which was successfully validated experimentally with high enantiomeric excess and anti selectivity (70% yield, >99% ee, anti :syn = 94:6).

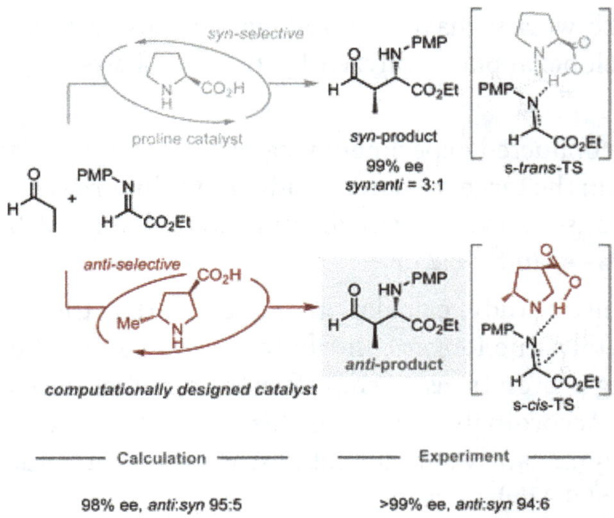

Fig. 10.3 Computational design of organocatalysts for enantio- and anti-selective Mannich reactions; calculations were performed at the HF/6-31G* level.

While the general approach for designing proline catalysts involves introducing a bulky substituent at the C2 position to enhance their enantiodiscrimination ability, this work presents a different strategy (using a small substituent) guided by quantum chemical calculations, highlighting the potential of computational approaches for catalyst design in the development of new methodologies.

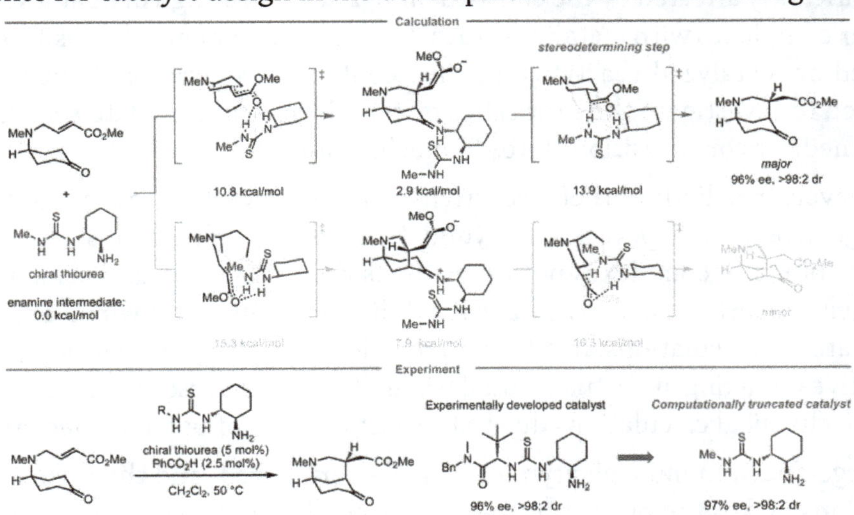

Fig. 10.4 Computational truncation of a thiourea catalyst for the asymmetric desymmetrization of cyclohexanes; calculations were performed at the M06-2X/6- 311+G(d,p)/CPCM(CH2Cl2) level.

In addition to designing catalysts for superior catalytic activity, computational methods can also facilitate the simplification of catalyst structures. In this context,

Paton, Dixon, and co-workers have reported the enantioselective desymmetrization of a prochiral cyclohexanone catalyzed by thiourea-based organocatalysts (Fig. 10.4).

Initially, they conducted experimental catalyst screening trials to enhance the enantioselectivity in the target reaction, leading to a thiourea catalyst bearing both tert leucine and trans-1,2-cyclohexanediamine units that furnished the product in 84% yield with 96% ee, and >98:2 dr.

In the mechanistic study, calculations indicated that the tert-leucine moiety on the thiourea catalyst no longer contributes to controlling the selectivity at the stereo determining proton transfer step after the enantioselective intramolecular Michael addition. Accordingly, replacing this unit with a methyl group resulted in a structurally significantly simpler catalyst with similar catalytic activity (83% yield, 97% ee, and >98:2 dr).

Computational design of transition–metal complexes

Transition-metal complexes have been studied extensively in the development of synthetic methodologies because they exhibit unique abilities that cannot be achieved using organic compounds alone, which has led to the discovery of new elementary processes and catalytic reactions that serve as valuable synthetic tools in organic synthesis. Given that the reactivity and selectivity of metal complexes are significantly affected by the constituent ligands, the design of ligand structures and their complexes with metals is critical for the development of transition-metal-mediated or -catalyzed reactions. In this context, computational chemistry has been used to investigate the properties of complexes and elucidate the structures of intermediates or transition states in mechanistic studies.

However, predicting their reactivity or selectivity poses a considerable challenge due to many factors beyond the cost of calculations. Altering the structure of metal complexes not only affects their reactivity and selectivity, but also other properties, such as catalyst stability and solubility, which are difficult to estimate via calculations. In addition, reactions are often influenced by solvents or additives, making it virtually unfeasible to fully understand their role using quantum chemical calculations due to the vast number of possible mechanisms.

Charge and spin multiplicity must also be considered, and the involvement of counterions makes the prediction even more challenging. Even when calculations are performed to account for the factors mentioned above, the estimation of energy is significantly affected by the calculation method, and the appropriate calculation level for each reaction is not yet well understood.

For example, calculations of the dispersion effect are highly dependent on the choice of functional. Due to the aforementioned considerations, predicting transition-metal-mediated or -catalyzed reactions is significantly more complicated than predicting organic reactions.

Despite these challenges, in certain cases, computational predictions can effectively aid in the development of organometallic reactions. For example, Schoenebeck and co-workers have reported an approach that uses computational methods for ligand design in the reductive elimination of benzotrifluoride from palladium complexes (Fig. 10.5). Previously, this reaction had been studied experimentally using bidentate phosphine ligands.

The palladium complex with Xantphos is able to effectively promote this step at 80 °C, while the complex with 1,2-bis(diphenylphosphino)ethane (DPPE) does not furnish the desired elimination product, not even at 130 °C. This result suggests that the large bite angle of the bidentate phosphine ligand plays a crucial role in the acceleration of reductive elimination.

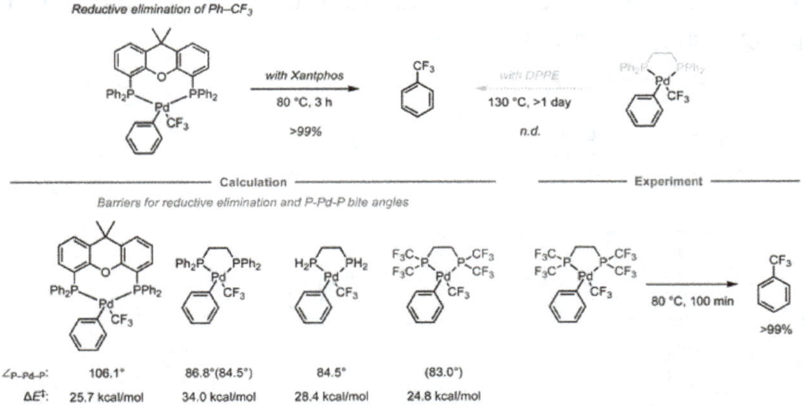

Fig. 10.5 Computational design of phosphine ligands for the reductive elimination of benzotrifluoride from Pd complexes; calculations were performed at the ONIOM(B3LYP/6- 31+G(d) (with LANL2DZ for Pd): HF/LANL2MB) level. The bite angles shown in parentheses were calculated at the B3LYP/6-31+G(d,p) (with LANL2DZ for Pd) level.

Later, the authors also investigated the energy barriers for these reductive eliminations using DFT calculations.81 The computations revealed that the reaction with Xantphos had a much lower energy barrier compared to that with DPPE (barrier of electronic energy, DE‡ = 25.7 kcal mol-1vs. 34.0 kcal mol-1; bite angle ĐP–Pd–P = 106.1° vs. 86.8°).

In contrast, an analogous ligand with two hydrogen atoms on the phosphine atoms on DPPE instead of phenyl groups exhibited a significant reduction of these values (DE‡ = 28.4 kcal mol-1, ĐP–Pd–P = 84.5°), suggesting that an appropriate substituent on the phosphine atom might accelerate this step even when using an ethylene-bridged scaffold, beyond the effect of the bite angle.

Accordingly, replacing the phenyl groups in DPPE with trifluoromethyl groups resulted in a reduced barrier that was even lower than that with DPPE or Xantphos (DE‡ = 24.8 kcal mol-1, ĐP–Pd–P = 83.0). This observation was further validated experimentally, resulting in a quantitative conversion to provide benzotrifluoride at 80 °C. The acceleration in this case can be attributed to two factors, i.e., the destabilization of the complex caused by electrostatic repulsion between the trifluoromethyl groups of one of the phosphine ligands and the ligand bound to palladium, and the reduced electron density of the metal center, which often facilitates the reductive elimination step in other organometallic reactions.

Reactive metal complexes often cause decomposition or side reactions that can be problematic in catalytic reactions. Therefore, a balance between the reactivity and stability of metal complexes species is critical for reaction efficiency. In this context, Buchwald and co-workers reported the computational design of copper complexes with sufficient reactivity and stability to act as catalysts in the C–N cross-coupling of aryl bromides and amines (Fig. 10.6).

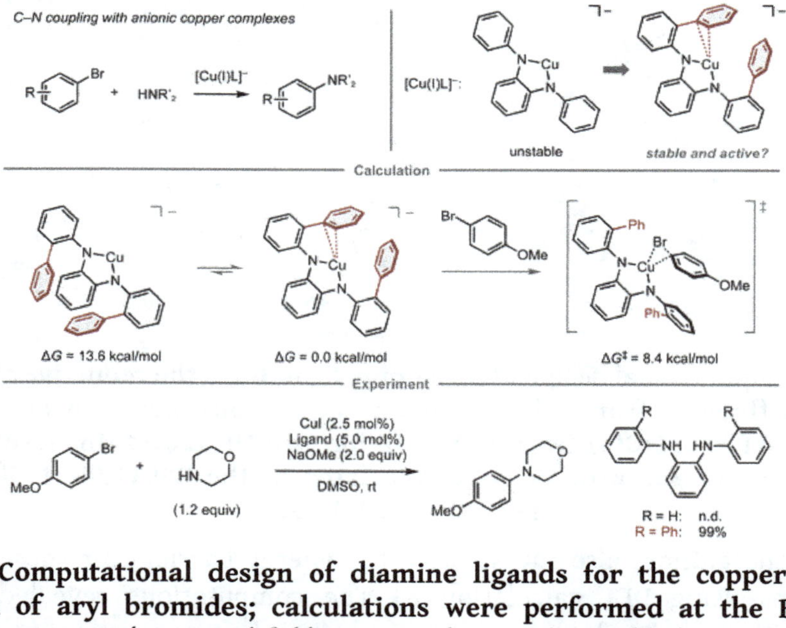

Fig. 10.6 Computational design of diamine ligands for the copper-catalyzed amination of aryl bromides; calculations were performed at the B3LYP-D3/ LACV3P/cc-pVTZ(-f)//B3LYP D3/LACVP/6-31G(d,p) level.

The copper complexes comprise copper(I) species and an anionic N1,N2-diarylbenzene 1,2-diamine ligand designed to accelerate the oxidative addition step of aryl electrophiles, which is conventionally problematic in copper-catalyzed cross-coupling reactions. Initially, the reaction of 4-bromoanisole and morpholine using a copper complex with an anionic N1,N2-diphenylbenzene-1,2-diamine ligand was investigated, albeit the desired amination product was not obtained, presumably due to the instability of the naked anionic copper species.

To this end, a computational analysis was conducted, which revealed that the introduction of two phenyl groups onto the ligand stabilizes the complex through copper– ð interactions (DG = 13.6 kcal mol-1), while maintaining sufficient reactivity for the oxidative addition of 4-bromoanisole (DG‡ = 8.4 kcal mol-1).

The reaction with this ligand proceeds at room temperature and provides the amination product in quantitative yield. While this work successfully demonstrates a predictive strategy for the development of catalytic reactions by focusing mainly on one elementary step, the prediction of entire catalytic cycles requires considering all individual steps in the catalytic cycle as well as the turnover for their catalytic efficiency.

Computational prediction of the key step in complex natural product syntheses

The computational prediction of chemical reactions is also potentially useful for retrosynthetic analysis, a concept used to devise synthetic routes toward complex molecules from readily available starting materials by converting a target compound to synthetic precursors and applying this process to each precursor, especially in the field of natural product synthesis.

Since Corey's report on computational retrosynthetic analysis, this approach has received considerable attention, leading to studies using advanced chemoinformatics methods and databases that contain experimental data pertaining to molecular synthesis. These studies enable innovative computational synthetic planning for the synthesis of complex molecules, which has been verified experimentally.

Given that complex natural product synthesis often involves key synthetic steps, a considerable number of experimental investigations are often required to identify suitable substrates and reaction conditions. Ultimately, unsuccessful results force modifications onto the original synthetic route. To avoid such dead ends and detours, quantum chemical calculations have been used for the prediction of key steps and prospective substrates, thus enhancing the reliability of the retrosynthetic plan.

In this context, Overman and co-workers reported the total synthesis of (–)-chromodorolide B, demonstrating the utility of computational predictions in a key step to achieve the desired stereoselectivity (Fig. 10.7).

In their first report, one of the key steps was a radical cascade reaction involving an intermolecular C–C bond formation and subsequent stereoselective 5-exo-trig cyclization. Although their developed conditions afforded the product, the yield and stereoselectivity were low, i.e., the major product consisted of the undesired stereoisomer. To gain insight into the structure of the transition state and improve the selectivity, they performed DFT calculations for the radical cyclization step.

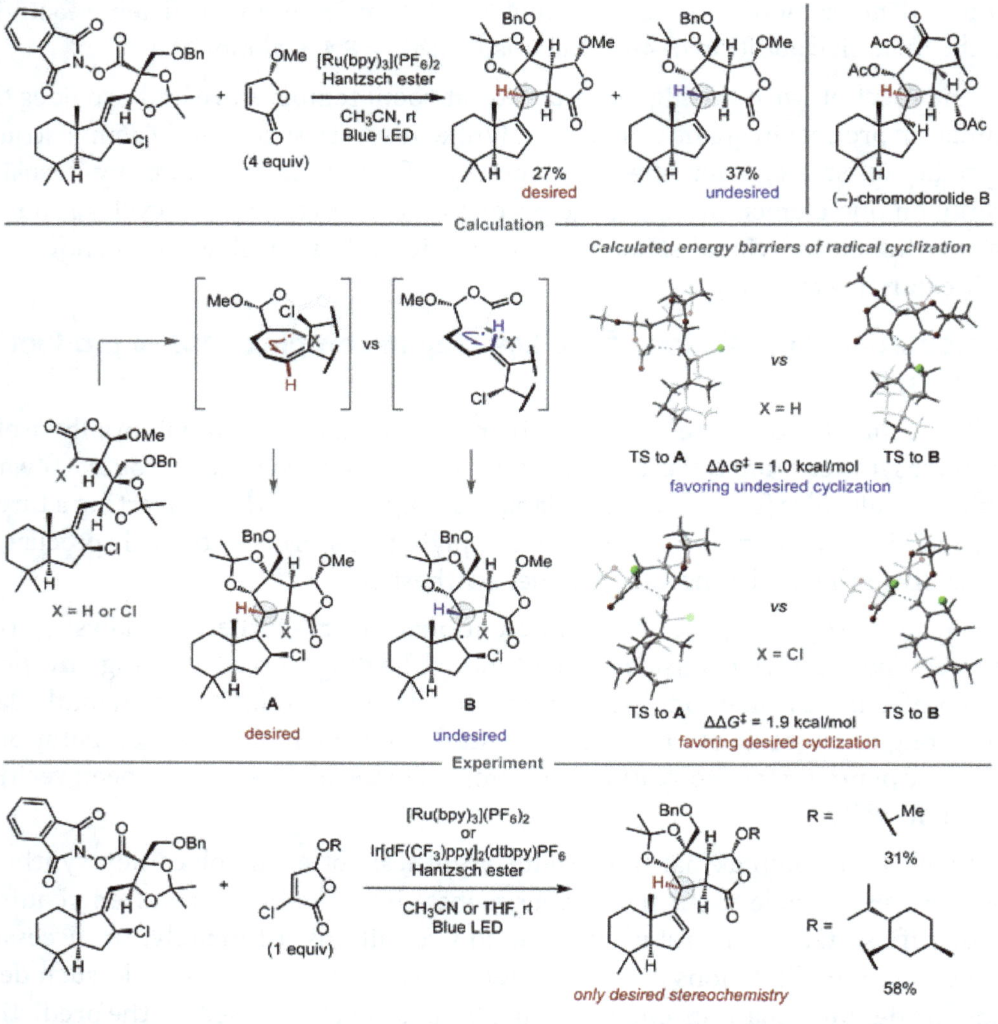

Fig. 10.7 Computational analysis of stereoselective radical cyclization toward the total synthesis of (–)-chromodorolide B; calculations were performed at the TPSSh-D3/Def2- TZVP/COSMO(CH2Cl2)//TPSS-D3/Def2-TZVP/COSMO(CH2Cl2) level.

While the alkyl radical derived from the original substrate exhibited the undesired preference (DDG‡ = 1.0 kcal mol-1, favoring the undesired stereochemistry B), consistent with the observed selectivity in the experiment, the cyclization of the chlorinated variant exhibited inverse selectivity, resulting in the desired stereoselectivity (DDG‡ = 1.9 kcal mol-1, favoring the desired stereochemistry A).

These distinct selectivities were rationalized in terms of the steric interactions between the substituent at the á-position of the ã-butyrolactone and the alkene moiety at the transition states of 5-exo-trig cyclization. In the case of chlorinated substrates, the steric effects govern the olefin geometry of the transition state towards the desired diastereomer.

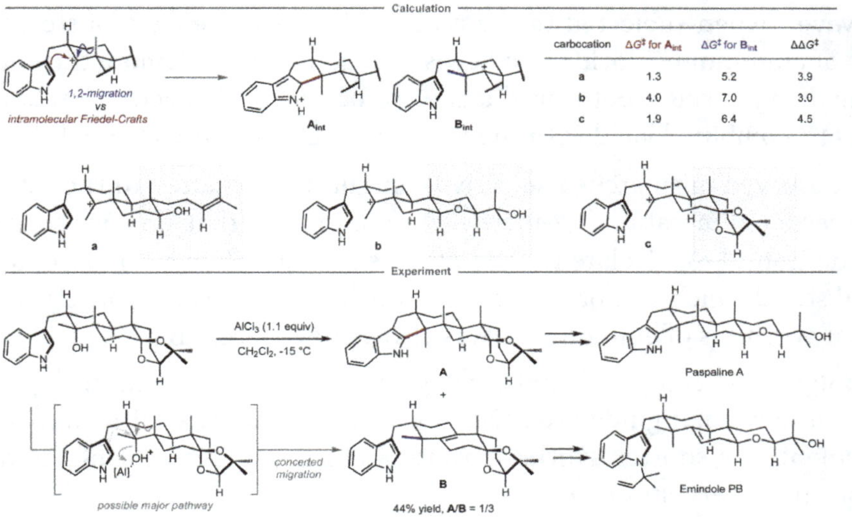

Fig. 10.8 Computational substrate design for the divergent synthesis of paspaline A and emindole PB. The calculations were performed at the mPW1PW91/6-31+G(d,p)//B3LYP/6- 31G(d) level.

Based on this notion, the reaction was performed using chlorinated substrates bearing either a methyl or a menthyl group, which resulted in selective cyclization and provided the desired product without the undesired diastereomer.

The introduced chlorine atom was removed under the applied reaction conditions. With this improved key step, the authors ultimately completed the synthesis of (-)- chromodorolide B with a higher overall yield compared to that of the first-generation route.

Quantum chemical calculations have also been applied for selecting prospective substrates even prior to experimental investigations. Newhouse and co-workers reported the total synthesis of paspaline A and emindole PB using computation-directed substrate selection (Fig. 10.8).

In that study, the authors aimed to synthesize both compounds from the same precursor in a divergent fashion. To this end, calculations were conducted to assess the energy barriers of the Friedel–Crafts type cyclization and the 1,2-methyl shift leading to each key scaffold, in which these pathways were explored from a carbocation intermediate bearing various substituents distant from the reaction site.

Calculations revealed that among carbocations a–c, the reaction of carbocation c has the highest preference for forming the hexacyclic motif in paspaline A. Inspired by these theoretical results, the synthesis commenced with the preparation of the computationally proposed precursor for carbocation c bearing a hydroxy group as a leaving group.

However, when subjected to the reaction in the presence of a stoichiometric amount of aluminum chloride as a Lewis acid additive to promote the elimination of the hydroxy group, both products were obtained via cyclization and methyl shift in 44% combined yield with the inverse selectivity ratio (A/B = 1/3).

The observed mismatched selectivity might be attributed to the involvement of a concerted elimination/1,2-methyl shift, a process that was not considered in the calculations. From isolated A, the synthesis of paspaline A was accomplished in 9 overall steps, which is shorter than the previously reported synthetic route. The first total synthesis of emindole PB was also achieved from B.

Although this computational strategy was unable to accurately predict the experimental ratio, it guided the selection of substrates that ultimately led to the development of a shorter synthetic route and the achievement of the first total synthesis in a divergent manner.

RECENT DISCOVERIES AND BREAKTHROUGHS IN REACTIVITY RESEARCH

A new era of space exploration

Need to be reminded of how incredibly vast our universe is? The first ever photos from the James Webb Space Telescope are awe-inspiring. While this is the most technically advanced and powerful telescope ever created, the learnings about our universe will lead to future missions and exploration for generations ahead. Recently, the newest mission to the moon was launched as NASA's Artemis Program which will pave the way for a future mission to Mars. This new era of space exploration will drive technological advancements in fields beyond astronautics and stimulate progress in real-world applications like materials, food science, agriculture, and even cosmetics.

Amilestone in AI predictions

For decades, the scientific community has chased a greater understanding of relationships between protein functions and 3D structures. In July 2022, Deep Mind revealed that the folded 3D structure of a protein molecule can be predicted from its linear amino-acid sequence using AlphaFold2,

RoseTTAFold, and trRosettaX-Single algorithms. The algorithms' predictions reduced the number of human proteins with unknown structural data from 4,800 to just 29. While there will always be challenges with AI, the ability to predict protein structures has implications across all life sciences. Key challenges in the future include modeling proteins with intrinsic disordered properties and those that change structures by post-translational modifications or to environmental conditions. Beyond protein modeling, AI advancements continue to reshape workflows and expand discovery capabilities across many industries and disciplines.

Developing trends in synthetic biology

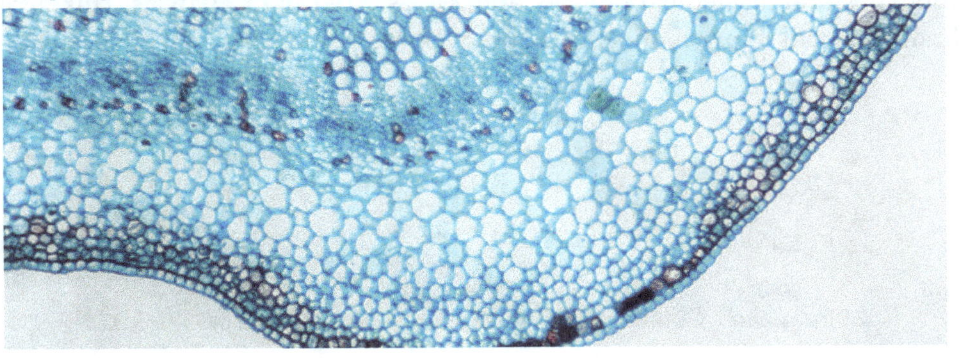

Synthetic biology has the potential to redefine synthetic pathways by using engineered biological systems (i.e., microorganisms, for which a large part of the genome or the entire genome has been designed or engineered) to manufacture a range of biomolecules and materials, such as therapeutics, flavors, fabrics, food, and fuels. For example, insulin could be produced without pig pancreas, leather without cows, and spider silk without spiders. The potential in life sciences alone is unbelievable, but when applied to manufacturing industries, synthetic biology could minimize future supply chain challenges, increase efficiency, and create new opportunities for biopolymers or alternative materials with more sustainable approaches. Today, teams use AI-based metabolic modeling, CRISPR tools, and synthetic genetic circuits to control metabolism, manipulate gene expression, and build pathways for bioproduction. As this discipline begins to cross over into multiple industries, the latest developments and emerging trends for metabolic control and engineering challenges are showcased in a 2022 Journal of Biotechnology article.

Single-cell metabolomics set to soar

While much progress has been made in genetic sequencing and mapping, genomics only tells us what a cell is capable of. To have a better understanding of cellular functions, proteomic and metabolomic approach es offer different angles for revealing molecular profiles and cellular pathways. Single-cell metabolomics gives a snapshot of the cellular metabolism within a biological system. The challenge is that metabolomes change rapidly, and sample preparation is critical to understand cell function. Collectively, a series of recent advancements in single-cell metabolomics (from open-sourced techniques, advanced AI algorithms, sample preparations, and new forms of mass spectrometry) demonstrates the ability to run detailed mass spectral analyses. This allows researchers to determine the metabolite population on a cell-by-cell basis, which would unlock enormous potential for diagnostics. In the future, this could lead to the ability to detect even a single cancerous cell in an organism. Combined with new biomarker detection methods, wearable medical devices and AI- assisted data analysis, this array of technologies will improve diagnosis and lives.

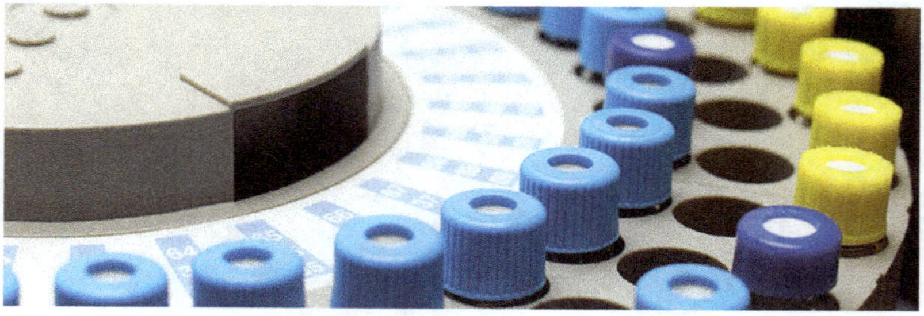

New catalysts enable greener fertilizer production

Every year, billions of people depend on fertilizers for the ongoing production of food, and reducing the carbon footprint and expenses in fertilizer production would reshape the impact agriculture has on emissions. The Haber-Bosch process for fertilizer production converts nitrogen and hydrogen to ammonia. To reduce energy requirements, researchers from Tokyo Tech have developed a noble-metal-free nitride catalyst containing a catalytically active transition metal (Ni) on a lanthanum nitride support that is stable in the presence of moisture. Since the catalyst doesn't contain ruthenium, it presents an inexpensive option for reducing the carbon footprint of ammonia production. The La Al-N support, along with the active metals, such as nickel and cobalt (Ni, Co), produced NH3 at rates similar to conventional metal nitride catalysts. Learn more about sustainable fertilizer production in our latest article.

Advancements in RNA medicine

While the application of mRNA in COVID-19 vaccines garnered lots of attention, the real revolution of RNA technology is just beginning. Recently, a new multivalent nucleoside-modified mRNA flu vaccine was developed. This vaccine has the potential to build immune protection against any of the 20 known subtypes of influenza virus and protect against future outbreaks. Many rare genetic diseases are the next target for mRNA therapies, as they are often missing a vital protein and could be cured by replacing a healthy protein through mRNA therapy. In addition to mRNA therapies, the clinical pipeline has many RNA therapeutic candidates for multiple forms of cancers, and blood and lung diseases. RNA is highly targeted, versatile, and easily customized, which makes it applicable to a wide range of diseases. Learn more about the crowded clinical pipeline and the emerging trends in RNA technologies in our latest CAS Insight Report.

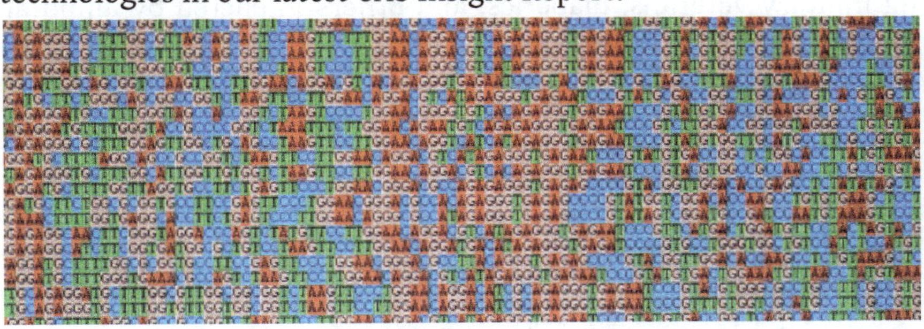

Rapid skeletal transformation

Within synthetic chemistry, the challenge of safely exchanging a single atom in a molecular framework or inserting and deleting single atoms from a molecular skeleton has been formidable. While many methods have been developed to functionalize molecules with peripheral substituents (such as C-H activation), one of the first methods to perform single-atom modifications on the skeletons of organic compounds was developed by Mark Levin's group at the University of Chicago. This enables selective cleaving of the N–N bond of pyrazole and indazole cores to afford pyrimidines and quinazolines. Further development of skeletal editing methods would enable rapid diversification of commercially available molecules, which could lead to much faster discoveries of functional molecules and ideal drug candidates.

Advancing limb regeneration

Limb loss is projected to affect over 3.6 million individuals per year by 2050. For the longest time, scientists believed the single biggest key to limb regeneration is the presence of nerves. However, work done by Dr. Muneoka and his team demonstrated the importance of mechanical load to digit regeneration in mammals and that the absence of a nerve does not inhibit regeneration. The advancement of limb regeneration was also achieved by researchers at Tufts University who have used acute multidrug delivery, via a wearable bioreactor, to successfully enable long-term limb regeneration in frogs. This early success could potentially lead to larger, more complex tissue re-engineering advances for humans, eventually benefiting military veterans, diabetics, and others impacted by amputation and trauma.

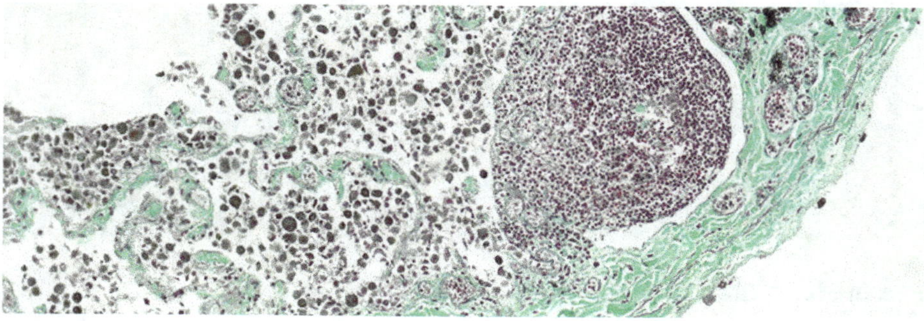

Nuclear fusion generates more net energy with ignition

Nuclear fusion is the process that powers the sun and stars. For decades, the idea of replicating nuclear fusion on earth as a source of energy, in theory, could fulfill all the planet's future energy needs. The goal is to force light atoms to collide so forcefully that they fuse and release more energy than consumed. However, overcoming the electrical repulsion between the positive nuclei requires high temperatures and pressures. Once overcome, fusion releases large amounts of energy, which should also drive the fusion of nearby nuclei. Previous attempts to initiate fusion used strong magnetic fields and powerful lasers but had been unable to generate more energy than they consumed.

11

CHALLENGES AND FUTURE DIRECTIONS

CURRENT CHALLENGES IN UNDERSTANDING CHEMICAL REACTIVITY

Synthesis & Manufacturing

Chemist wants to understand not only the substances and transformations that occur in the natural world, but also those that are permitted by natural laws. Thus the field involves both discovery and creation. They want to discover the components of the chemical universe – from atoms and molecules to organized chemical systems such as materials, devices, living cells, and whole organisms and want to understand how these components interact and change as a function of time. They also consider the unknown molecules & substances and interactions that could exist.

Science has become increasingly interdisciplinary. Investigating a single compound, a single reaction or a single process may fall within the expertise of a single discipline or subdiscipline but the situation is different when the investigations are extended to systems – full assemblages of related components that address the same function or to processes, where integrated systems of operations work in concert to produce a product. Understanding, developing, manipulating the systems and processes often require the synergistic advantages of the entire range of the chemical sciences – from fundamental chemistry, to chemical engineering, and advanced areas of science and technology, to create scientific understanding and benefit for society.

The goal of the basic science is to invent new types of transformations. Transformations are part of the universe of chemistry and they are the tools that make possible to create interesting and useful new substances. It is also important to invent overall strategies to achieve multistep syntheses involving a sequence of transformations. In synthesis and chemical manufacturing there is a general need to develop separation methods. There are two goals related to separations.

One is to learn how to isolate and purify the desired components with high efficiency, and the other is to develop synthetic processes that do not need such separation. For manufacturing, a further goal is to eliminate solvents or at least replace them with harmless inexpensive and easily removed or recovered solvents. The extra chemicals produced during chemical reactions must be removed and disposing of them adds cost and the potential for environmental problems. A synthetic chemist after recognizing the challenge looks for inspiration in several places for transformation that cannot be done with current methods. Synthetic inspiration can come from an attempt to make drastic improvements in the properties of existing substances. For example, some metals and metal compounds become superconducting at extremely low temperatures.

Challenges and Opportunities for the future

In spite of the impressive progress achieved by synthetic chemists in the past several decades, the state of the art of methods for chemical synthesis remains at an early stage of development. There is great need for new reactions and synthetic methods to permit any substance, of any complexity, be it organic or inorganic in structure, to be synthesized with high chemical efficiency, in a small number of steps and with minimal cogeneration of any waste products.

Synthesis and manufacture is important because the expansion of chemical universe through synthesis leads to great new scientific insights. The promise of better medicines and better materials depends on the ability of synthetic chemists to create new transformations and to use them in the creation and manufacture of new substances.

Chemical and Physical Transformations of Matter

A principal goal of the chemical sciences is to understand and manipulate chemical and physical transformations of matter. Transformation means a change in composition, state, or organization of matter, or movement or rearrangement of materials by flow, heat or diffusion. A chemical transformation can occur when molecules collide with sufficient energy. Generally a reaction occurs when the molecules are heated or irradiated to provide the energy necessary to overcome the activation barrier that separates reactants from products.

The detailed molecular structures involved throughout the entire procession from starting materials through the transition states to products in a chemical reaction are so far no directly observable, there are certain classical ways in which reaction mechanisms are effectively deduced. One route is through the use of kinetics, the observation of how the rate of a reaction depends on concentrations of various components, and on temperature and sometimes pressure.

Some chemical reactions, such as the change of diamond to graphite under normal conditions of temperature and pressure or the spontaneous conversion of left-handed to right-handed amino acids, are so slow that their rate of change is measured in millions of human lifetimes or longer whereas the photodissociation process is so fast that typically takes 10 to 100 femtoseconds. Thus Chemical scientists possess the tools to make kinetic measurements over the full range of reaction rates.

Super critical fluid processing is exploiting knowledge about this domain of phase behaviour for new applications in separations, chemical reactions and materials production. Supercritical carbon- dioxide is now used as a solvent in improved, environmentally benign methods for decaffeination of coffee and for dry cleaning.

The importance of physical transformations in making chemical products implies a vast opportunity for understanding structures of many kinds of important chemical products at higher levels and longer lengthscales than those of individual molecules. Chemists have synthesized a spectacular array of submicron – and nano particles with well-defined size and atomic structure and very special properties.

Transport phenomena play crucial roles in both the creation and performance of structural chemical products. It presents an important challenge in the progression toward smaller length-scales of microscale, nanoscale and molecular scale. Chemical processing will need to exploit small-scale reactors in new applications.

Challenges and Opportunities for the future

The femtosecond spectroscopy has made it possible to observe reactions on a short time scale, a major challenge is to devise ultrafast techniques that will permit observation of the actual molecular structure of a transition state, not just its rate of passage. As a general goal, we want to be able to make moving pictures of the reactions themselves, observing all the intermediate states and the rates at which they interconvert. Such moving pictures can be generated even now, by computer simulation of the reaction: the problem is to determine whether those pictures are correct. Thus a second challenge is to interface with theoretical chemistry in getting the best possible calculations, and then devise experimental tests of the major theoretical predictions to see whether experiments confirm the correctness of the calculations.

Much work has been done to understand how metal ions react or catalyze the reactions in solution. There is still need to understand how the enzymes work.

When we do understand them in detail, we should be able to produce biomimetic catalysts for useful processes in manufacturing.

Isolating, Identifying and Measuring substances

Chemical scientists want to explore the natural world and identify all its chemical components. They also want and need to identify all the new chemical substances produced directly and indirectly as a result of their synthetic and manufacturing endeavors. Detecting known substances and determining their quantity is also important. In synthetic research, it is essential to know the relative proportions of various reaction products. In manufacturing it is important to detect any impurities in the product and to determine whether they are present in a significant amount.

The frontiers in this field lie in improving sensitivity to detect vanishingly small quantities, to separate extremely complex mixtures of chemical substances and to assess the structures or compositions of components. Measurements using very small or very dilute samples can present a major challenge in clinical medicine or environmental analyses. A vital activity of the chemical sciences is the determination of structure. Some molecules exist in more than one structure (conformation). Some molecules are so floppy that structural characterizations really refer to averages among several structures. Yet other molecules are sufficiently rigid that molecular structure can be quite precisely determined. The techniques available to achieve molecular structure determinations are limited eg. X-ray diffraction technique, microwave spectroscopy, nuclear magnetic resonance spectroscopy etc.

A major limitation of diffraction techniques has been the need to obtain crystalline samples. If scientists could learn how to crystallize large molecules in a routine manner, a breakthrough would result. The "weighing" of a molecule of a chemical substance and of its fragments has great utility in both assessing molecular identity and determining molecular structure. Mass spectrometry requires that the material being studied be converted into a vapour. The invention of new ways to volatize molecules, from solids and from surfaces, has revolutionized and re-invigorated this field.

Challenges and Opportunities for the future

Chemical measurements will continue to be challenged by the broad advances of the chemical sciences. New fundamental knowledge is essential for advancement in the measurement sciences particularly in nanoscale science and technology, which are at the heart of future advances in automated storage and retrieval of information.

Chemical Theory and Computer Modeling

Computational chemistry and process systems engineering play a major role in providing new understanding & development of computational procedures for the simulation, design & operation of systems ranging from atoms and molecules to industrial scale processes.

Science proceeds by both experiment & theory. Simple experimental facts without a theory to interpret them do not satisfy our need for understanding. Theory validated by experiment runs through all of chemistry and almost all its branches now use computers. Structures are determined by computer treatments of x-ray data, potential new drugs are analyzed by computer modeling and even synthetic strategies to make a desired target molecule are developed using computer programs created for the purpose. Modeling and simulation are extremely important tools in the chemical sciences.

The understanding of complex chemical processes depends on modeling and computation. Recent advances in computing not only have enabled more accurate & reliable calculations, but they have also provided new tools for interpreting the output of the calculations. Modern computer graphics i.e. molecular graphics, simulations and animations have greatly enhanced the ability of scientists to understand & utilize the results of their computations.

Challenges and Opportunities for the future

Improvements in DFT functionals will improve the accuracy of DFT calculations. One specific challenge will be to derive density functional theory from fundamental theory in a way that reveals how to incorporate successive approximations for the exchange & correlation terms. Some work on high level methods with better scaling may allow more immediate extension of high accuracy calculation to larger systems. We need to develop new & powerful computational methods that span from the atomic and molecular level to the chemical process & chemical enterprise level in order to allow their effective integration for multiscale simulation and optimization.

Interface with Biology and Medicine

The goal of fundamental science at the interface of chemical sciences & biology is to understand life in chemical terms. The mechanisms of life are revealed when we understand their molecular details. Moreover, molecular understanding of biology plays the major role in guiding drug discovery & molecular level understanding of biology is fundamental in developing diagnostic methods.

Molecular analysis of living cells has led to a flow plan of information in molecular biology, known as the Central Dogma: The information necessary for a cell is encoded in the double helix of DNA. Many of the computational techniques are related to recent advances in sequencing of DNA.

Bioinformatics includes computational chemistry with the goals of predicting function and three dimensional shape directly from the amino acid sequence-by comparison with sequence, function and structural information for other proteins. Knowing protein structure can provide direct benefits to human health. The three dimensional structural information derived from the X-ray crystal structure combined with computer modeling techniques allowed chemists to design potent, selective inhibitors of the protease enzyme.

Challenges and Opportunities for the future

The challenge is to discover the chemical identity of all the molecules that make up living organisms and the way they bind to each other and organize into biological structure such as nuclei, ribosome etc. Our driver for new discovery will be the completion of the human genome project. This project will help in determining the locations and sequences associated with the genes of the human genome. New and improved drugs will be required to fight disease and improve quality of life.

Materials by Design

A material is defined as the substance or substances out of which a thing is made or composed. The overall goal of materials research within the chemical sciences is to explore, design and control through synthesis and processing the relationships among structure, properties, composition that determine the useful behavior of all materials. A frontier challenge is to investigate the chemistry and properties of single isolated molecules and compare that behavior with the average molecular behavior in an assembly, solution or condensed phase of molecules. The traditionally made stable materials will last a long time and do not requires to be replaced. This longevity has the undesirable consequences of creating waste that requires significant energy to process or it clutters the landscape when discarded.

Challenges and Opportunities for the future

The science of materials reflects the way that chemistry has changed from a field concerned with atoms and molecules and their properties in isolation to a field concerned with organized interactive systems. This change opens opportunities and challenges in fundamental science to know the chemical components of the stuff and how do these components interact to produce the properties of the stuff. The solution of these questions will greatly add to understand the chemical universe.

Atmospheric and Environmental Chemistry

Geochemists have made progress in learning the chemistry of earth and its components including rivers, lakes and oceans. A field called earth systems engineering is emerging and this field will deal with global warming, carbon sequestration and environmentally benign manufacturing. It will also address new analytical, computational and assessment techniques through which global scale interaction in complex systems can be better understood and optimized.

The design, development and implementation of chemical processes and products to reduce or eliminate substances hazardous to human health and the environment is known as Green chemistry. This definition has been expanded to twelve principles.

To illustrate challenges and opportunities for green chemistry a simple acid catalysed reaction can be considered where Bronsted acids or Lewis acids are used. These reactions are common in chemical, petrochemical, pharmaceutical allied industries. These acids are cheap, readily available and very active but they are difficult to separate from the organic products and their use leads large volumes of hazardous waste.

In the petrochemical industries the reactions are common with solid acids. Solid acids including zeolites results relative low quantity of waste and can be removed from the products. The use of zeolites, clays mixed metal oxides, functionalized polymers and supported reagents as well as lanthanide triflates9 for liquid phase organic reactions and ionic liquids10 for some reactions stands solution for the problems in green chemistry.

Challenges and Opportunities for the future

Knowledge of complex chemistry by which the atmosphere and the earth react including the dependence of global climate on carbondioxide concentrations in the atmosphere is required. Scientist will need to investigate effective ways to trap or reduce the generation of carbondioxide that would otherwise build up in the atmosphere. The consequences of human action on the environment are not always foreseeable so to meet this challenge it is essential that we continue to cultivate a strong base of fundamental knowledge in the chemical and biological sciences and tools should be prepared not only to create new technologies but also to anticipate their consequences.

Energy for Future

Scientists should invent new ways to generate and transport energy for human use and provide for the needs and aspirations of a growing population in a sustainable manner. New ways should be discovered to minimize the energy used for human activities including manufacturing.

Challenges and Opportunities for the future

The challenges and opportunities in the field of energy are critical for a world in which inexpensive, readily available fossil fuels will eventually be exhausted. We need to devise better ways to use solar energy eg. By creating cells for conversion of sunlight to electricity, replacing combustion by fuel cell technology. We need to invent rechargeable batteries that are practical for vehicles that have electric motors instead of gasoline engines.

National and Personal Security

Chemical science and technology must contribute to the enhancement of national and personal security by providing fundamental understanding and new developments to defend against military, terrorist or criminal attack and give warning of accidental or natural disasters. The goal to which scientist can contribute begin with early detection and prevention of an attack or event but if prevention is not possible, they extend to mitigation of the effects and subsequent remediation of damage. Biological warfare agents present a threat because the original viruses or bacteria can multiply and infect additional people. New vaccines and new approaches for delivering drugs and vaccines are required to discover. The development of new vaccines will need to be carried out in full recognition that genetically modified pathogens could be used in an enemy attack.

Chemical warfare agents are extremely toxic and very fast acting. Scientists must develop better understanding of their mechanism of action and use this information to device possible remedies .Research by scientists will be required for the development of new analytical techniques to detect nuclear proliferation threats and treaty violations. This will require establishing the characteristic signatures of both production and testing of weapons. Detection of these signatures will depend on chemical spectroscopy techniques and advances in remote sensing.

EMERGING TRENDS AND TECHNOLOGIES

Technologies such as membranes, metal-organic frameworks and, more recently, hydrogels, have emerged as a practical possibility, which could enable applications such as floatable production platforms, with promising potential in large-scale production, sometimes straight from seawater, a sustainable source of hydrogen.

Dual-ion batteries

Electricity as we know it has a major flaw—it is surprisingly hard to store. To date, one of our best solutions are lithium-ion batteries, an advancement that was recently recognised with the 2019 Nobel Prize in Chemistry. During the past few decades, these devices have enabled the miniaturisation of energy-storage devices, currently used in laptop computers, mobile phones, and electric vehicles. Despite their high energy density, lithium-ion batteries still present some downsides. In fact, if you were to use state-of the-art batteries to power your house, you would need a device of over one ton to store enough energy for one week.

Moreover, the scarcity of lithium and cobalt limit future developments, and their links to conflict minerals clashes with SDG 12 on sustainable production patterns. Hence, newer devices such as dual-ion batteries (DIBs) have attracted the attention of the scientific community. Whereas in classic lithium-ion batteries, only cations move along the electrolyte, in DIBs both anions and cations participate in the energy storage mechanism. They also exhibit some fundamental differences in the cell setup—in DIBs the ions in the electrolyte are also active, which directly influences characteristics such as capacity and voltage.

DIBs could be an interesting alternative for grid storage applications. Their electrodes can be manufactured out of cheap and abundant materials using greener routes. Traditionally, the fabrication of lithium-ion batteries involves the use of toxic organic solvents. On the other hand, researchers envisioned fabricating DIBs using water processing, enhancing sustainability and reducing the cost. Although the first DIB prototypes also relied on lithium, now chemists have found new solutions that use sodium, potassium, or aluminium—all of which are copious and widely available worldwide. Discovered only a few years back, DIBs are still facing some challenges - researchers need to better understand their mechanism in order to improve their capacity, reversibility, and lifetime. Yet, industrial innovations are starting to blossom.

Recently, a team in China reported the first prototypes for viable pouch cells based on the dual-ion approach, and companies such as Ricoh and Power Japan Plus are already investing in this attractive technology. The latter has even tempted electric car maker Tesla with "a fully recyclable battery that charges faster than lithium while still allowing a range of 300 miles." In brief, DIBs present a series of advantages in terms of cost, lifetime, and sustainability that align with SDG 7. Besides, they provide a safer choice, according to experts. Since there is no intrinsic oxygen available in the cell, the 'fire tetrahedron' is incomplete, preventing accidental combustion.

Aggregation-induced emission

Nowadays, luminescent materials are ubiquitous: from LEDs to bio-imaging techniques. Since most of these substances usually feature a plethora of aromatic moieties, molecules tend to stack at high concentrations, which eventually kills luminescence. This effect is known as aggregation-caused quenching. In contrast, back in 2001 researchers observed the opposite phenomenon—certain luminogens showed very weak emission in diluted solutions and an intense emission when the molecules piled up. At first, the concept received very little attention, but now it has become a vast field of study. Aggregation-induced emission (AIE) has transformed the way people think about luminescence.

Molecular shapes are the key to understanding this effect. Unlike classic luminogens, the molecules that are AIE-active are non-planar. They are like miniature propellers, continuously moving. However, when they aggregate, rotation stops, and all of their energy is released in the form of light. Since the discovery of AIE, chemists have identified several families of compounds that exhibit this effect, including classic luminogens such as polyaromatic compounds and organometallic complexes, and more exotic products such as polymers, oligosaccharides, and nanoparticles.

AIE has opened new avenues in the development of luminescent materials—it has already found applications in OLED devices, sensors, and novel bio-imaging tools. The New York Times highlighted AIE's potential to reach the real-world very soon. In fact, start-ups that market AIE technology are blossoming—two good examples are AIEGEN Biotech, in Hong Kong, and Luminicell, in the US. The latter also sells their fluorescent nanoparticles for live cell tracking through leading chemical supplier Merck.

Microbiome and bioactive compounds

We contain multitudes. Over 10 trillion microbes live in our guts, respiratory track, and skin. Our microbiome may be modifying our behaviour, and research suggests it could also trigger diseases such as cancer, as well as determine our response to treatment. All of these bacteria constantly release metabolites in response to different stimuli in their environment. Chemistry could play a key role in screening and identifying all these different molecules, which could eventually be isolated and used as novel therapeutic candidates.

Very recently, a team in Princeton took this approach to the next level. Using different computational tools, they analysed bacterial genomes and identified gene clusters that codified the biosynthesis of small molecules. Then, they expressed these instructions in genetically-modified bacteria and obtained a series of molecules with strong antibacterial activity. Although this field—functional metagenomics—has progressed slowly in the past few decades, this new development has been characterised as "a game-changing approach, [with] the potential to revolutionise discovery." The microscopic life within us is immensely diverse. Chemists and biochemists may find a myriad of new bioactive compounds encoded in the genomes of bacteria, contributing directly to SDG 3. Understanding and unravelling the secrets of our microbiome could revolutionise the future of healthcare.

Liquid gating technology

The idea of using liquids as a structural material to build responsive gates sounds counterintuitive—even verges on science fiction. However, this idea, originally proposed in 2015, has already become a reality, and could soon bring many novel applications. Usually, liquid membranes work thanks to differences in concentration and potential astride the border. However, liquid gating membranes respond to pressure changes that rely on capillarity. At the microscale, phenomenon allows certain liquids to selectively open and close pores on-demand.

Liquid gates can selectively process mixtures of fluids without clogging. Thus, researchers predict that they could become extremely useful for large-scale filtration and separation processes. Among other things, liquid gates could accelerate the progress towards SDG 6, which looks to ensure access to clean water and sanitation for all. Moreover, since liquid gates require no electricity at all, they ensure huge energy savings.

Besides separation, liquid gates could find uses in many other fields, such as chemical sensors, microfluidic arrays, high efficiency catalysis, biological tissue printing, and lab on-chip applications. Despite its novelty, liquid gating technology has already been recognised as a key innovation by prospecting company TechConnect, in the U.S. "It is sufficiently mature and attractive for licensing and investment," they said. Hopefully, liquid gates will soon be scaled-up and adopted by key players in the chemical enterprise.

High-pressure inorganic chemistry

We all perform differently under pressure. Chemicals are no exception, and the most exceptional phenomena take place at extreme conditions. For instance, researchers have squeezed benzene into superstrong, ultra-thin diamond nanothreads and recently provided spectroscopic evidence of having prepared metallic hydrogen. High-pressure science is no longer a niche field. The newest technological advances allow to closely monitor samples under high-pressure environments, enhancing our understanding of materials.

These experiments involve pressures of up to 500 GPa - equivalent to five million times the average atmospheric pressure. To reach these immense strengths, scientists need to trap their samples between two diamond tips, which is commonly known as a diamond anvil cell. Further enhancements, such as combining diamond anvils with high energy X-rays, allow even higher pressures, reaching limits around 640 GPa.

Under ultra-high pressure, the rules of chemical bonding reshape. Stoichiometry laws blur—researchers have isolated 'cousins' of common salt from Na3Cl to NaCl7. In addition, some compounds that are undoubtedly inert at ambient conditions suddenly become reactive. Traditionally inactive species, such as dinitrogen, carbon monoxide, and carbon dioxide polymerise under extreme pressure and temperature, yielding products that, in some cases, survive depressurization and can be isolated at atmospheric pressure. High pressure also enhances well-known effects such as luminescence and superconductivity.

Chemistry gets very complex in these conditions, but at the same time it gets really interesting. Discerning the transformations that take place under ultra-high pressures could lead to new molecular species and novel materials with unprecedented properties, such has room temperature superconductivity or superhardness. Moreover, some of the knowledge acquired could be translated to room pressure processes— researchers hope to open new frontiers in chemistry.

Macromonomers for better plastic recycling

2020 marks the 100 anniversary of Hermann Staudinger's prestigious manifesto on polymerisation. Chemistry played a key role in the development of artificial polymers— durable and versatile materials that transformed our civilisation. However, said durability has turned against us: the building blocks of the twentieth century are now everywhere, accumulating in our landfills and polluting our oceans. Some experts predict that by 2050, the total amount of plastic in the oceans will weigh more than the total amount of fish. Now, chemists must find a solution.

Many research groups are looking into more efficient ways to recycle the polymers we know, as reflected by last year's IUPAC Top Ten Emerging Technologies in Chemistry. Additionally, other groups are investigating new types of polymers that can be easily recycled. Solutions include plastics that break down upon exposure under UV light, and macromolecules bearing responsive "end caps" that trigger depolymerisation on-demand.

Redesigned monomers and macromonomers are an up-and-coming strategy to craft more sustainable plastics. Chemists rely on radical ring-opening reactions, which allow them to incorporate heteroatoms and functional group—such as ester—in structures that, traditionally, have an all-carbon backbone.

The resulting polymers are easier to hydrolyse and recycle. Recently, several groups have optimised this technology, delivering a broad-range of biodegradable plastics that keep the attractive characteristics of conventional polymers. Starting from a widely available lactone, researchers developed a strong and stable polymer that can be recycled again and again under mild conditions.

These methods are far from being widely adopted. Nevertheless, chemists are moving in the right direction—re-thinking polymers and designing structures that ensure recyclability. Undoubtedly, chemistry is our best chance to find a solution to the plastic problem, contributing to at least five different SDGs at the same time.

Artificial intelligence

Artificial intelligence is transforming our society. Its market value is growing exponentially as it finds uses in finance, justice, transportation, and even healthcare. Chemistry is no exception. Researchers train algorithms to speed-up structure elucidation, enhance retrosynthetic analyses, design optimised reaction sequences, discover new drugs, and even run futuristic robotic laboratories. The possibilities are endless. "In the future, we will forget chemists used to be humans," believes chemist and inventor Lee Cronin.

The applications of artificial intelligence in chemistry are only just beginning—the biggest leaps in progress are yet to come. Researchers predict that these technologies have a tremendous potential. Among other things, they expect chemical reactions will become more reproducible, more easily scalable and eventually, greener and more efficient. Thanks to the combination of high-throughput methodologies and automated analyses, chemists could control and accelerate serendipity—turning accidental discoveries into a thorough and carefully-planned search. All these strategies could expedite scientific breakthroughs and solve increasingly-sophisticated problems.

Algorithms could also tackle broader issues. For instance, machines can systematically analyse the scientific literature and learn from virtually every piece of data ever published. This could not only help us recognise trends, but also identify possible solutions to bigger challenges related to energy, climate change, environment, and health. In fact, recent studies suggest that artificial intelligence has a positive impact towards achieving the SDG—enabling the accomplishment of 134 targets across the goals.

Technology will upgrade our role as chemists. Rather than replacing us, artificial intelligence will enhance chemical discoveries while freeing us from mundane and repetitive tasks. Thus, we will focus on creativity, enabling leaps only limited by our imagination.

Nanosensors

Sensors detect changes in the environment. In chemistry, the process of sensing involves two steps—recognition, when analyte molecules meet their receptor; and transduction, the 'translation' of that event into an output signal. Nanosensors work in a similar way, only they use nanomaterials as the active element. Chemical nanosensors are used in a myriad of applications, from monitoring pollution and food quality control to security and healthcare.

The field of sensors has progressed to the point of detecting single molecules. This has been dubbed "the ultimate sensitivity," and is paramount in healthcare applications, where detecting a single entity can be a matter of life or death.

Single-molecule sensing provides additional benefits, such as easily measuring heterogeneity in a sample or enabling calibration-free measurements. Experts believe that these technologies could be "paradigm changing".

Advances in chemistry and materials science have allowed significant improvements. Researchers have explored a tremendous variety of nanostructured materials—metal, oxides, carbon nanotubes, graphene, polymers—that, thanks to their high surface to volume ratio, provide significant benefits for sensing. Nanosensors are used across a number of fields in analytical chemistry. Recently, antibodies have attracted a great deal of attention due to the pandemic caused by coronavirus SARS-CoV-2.

Luckily, chemists have used the unique properties of nanomaterials to create antibody nanosensors that are extremely sensitive and specific. Gold nanoparticles, for instance, enable the detection of SARS-CoV-2 in under 15 minutes.

We are surrounded by sensors—they are inside our phones, fitness bands, smartwatches, and computers. Nanosensors will become increasingly popular, helping us distinguish fresh food from products about to expire, or upgrading our ability to

detect previously unknown brainwaves, unlocking potential treatments for diseases such as epilepsy. Sensors will help us better understand the world in which we live.

RNA vaccines

Vaccines prepare our immune system to fight diseases. Through different agents, vaccines induce the production of antibodies, molecules that recognise and trigger the destruction of pathogens. In particular, RNA vaccines have a very clever approach to this goal—the patient is administered an RNA sequence encoding the production of antigens, which eventually stimulate an immune response and the synthesis of antibodies. Although RNA vaccines have not been approved as yet for human use, they have shown promising results in clinical trials. Their potential to supply a speedy solution to prevent the infection caused by the novel SARS-CoV-2 coronavirus has again put them under the spotlight.

One of the advantages of RNA vaccines is that their synthesis can be easily scaled up. To develop classic vaccines, researchers need to grow the infectious agent in a cell culture, which requires using high volume reactors and a significant amount of time. On the other hand, RNA strands can be synthesised using methodologies that have been optimised—and even automated—for decades. Moreover, RNA vaccines can be designed very quickly. Robin Shattock's team at Imperial College London developed a candidate vaccine against COVID-19 within two weeks of getting the virus' genomic sequence. The team is confident they could have preliminary results by next year. This is a true advantage when compared to classical vaccines, which usually need up to ten years of development— and an average investment of half a billion dollars—before they reach the market.

In addition to COVID-19, scientists are exploring the potential of RNA vaccines to prevent other infectious diseases such as Zika, rabies, HIV, the fl u, and even cancer. Studies show that RNA vaccines could stimulate an immune response against cancer cells, making them an attractive alternative for novel immunotherapy treatments. The Bill & Melinda Gates Foundation have invested an initial amount of $52 million for the further development of this technology.

Moreover, several companies are investigating RNA vaccines. Among them stand out CureVac and BioNTech, both in Germany, and Moderna in the U.S. All of them have adjusted their pipelines to investigate RNA vaccines against SARS-CoV-2, and are confident that they could upscale production if needed. RNA vaccines advance quickly— in fact, Moderna's candidate (mRNA-1273) was ready to start Phase III clinical trials on thirty thousand volunteers in mid-June.

While still young, the field of RNA vaccines will probably grow tremendously in the coming years—especially given how rapid and adaptable production is. Plus, if RNA vaccines against COVID-19 are successful and get fast-tracked into the market, this could further foster the advancement of the technology.

Rapid diagnostics for testing

Rapid diagnostic tests are chemical assays suitable for swift medical screening. They normally involve a series of easy-to-follow steps and provide results within a few minutes. Moreover, these tests seldom require additional equipment, facilitating their use in resource-poor settings. Probably the best-known example is the home pregnancy test, of which more than thirty-five million units are sold annually in the U.S. alone. There are also rapid tests to diagnose diseases such as malaria, aids, and the flu. Rapid tests work thanks to chemical reactions. Often, they use antibodies to detect the presence of antigens. Antibodies are linked to different types of probes that undergo a certain chemical reaction if the test is positive—this usually involves a colour change, making the interpretation of the results very straightforward.

The current COVID-19 pandemic has led to shortages of laboratory equipment to perform more thorough PCR tests. Thus, scientists around the globe have prioritised the development of rapid tests to detect SARS-CoV-2 and diagnose people suffering the disease this virus causes, COVID-19. Some of them rely on the detection of RNA strands rather than antigens, and deliver results in under half an hour. Pharmaceutical company Abbott, developed a COVID-19 test that allegedly uses loop-mediated isothermal amplification, which gives results in only five minutes. However, the latter requires some laboratory equipment.

At the moment, the World Health Organization (WHO) does not recommend the use of rapid diagnostic tests that detect antigens for COVID-19 patient care. So far, only three companies have received both an emergency use authorisation from the U.S. Food and Drugs Administration and the CE Mark from the European Commission—Autobio Diagnostics, CTK Biotech, and Hangzou Biotech. Hence, chemists need to race against the clock to develop a suitable alternative that can yield significant results in a timely manner.

CHEMISTRY FOR A SUSTAINABLE FUTURE

Chemistry provides us with an unlimited set of tools to reshape our world into one that is safer and promises a more sustainable future. From designing more efficient tests to developing a successful treatment, chemistry will be central to tackle the current COVID-19 pandemic, one of the most difficult challenges our society has faced in the last few decades. Furthermore, while we flatten the curve to stop the spread of the coronavirus and grant access to healthcare for those in need, we must remember other threats on the horizon, such as pollution, climate change, and circular economy. Innovation in the chemical sciences is essential to attain most of the SDG, the aspirational goals set by the United Nations to promote prosperity while protecting the planet. This aligns flawlessly with the main mission of IUPAC - to apply and communicate chemical knowledge for the greatest benefit of humankind and the world. This new edition of the "Top Ten Emerging Technologies in Chemistry" keeps the same spirit—promoting the fundamental role of chemistry to protect society and our planet.

POTENTIAL AREAS FOR FUTURE RESEARCH

Measurement

By enabling us to accurately and precisely ascertain the composition, structure, properties, and quantities of a variety of materials, measurement science fosters innovation across the chemical enterprise. State-of-the-art measurement technologies allow us to probe everything from subatomic particles to human health to ocean sustainability to our solar system, providing real-time or near-real-time data even when looking back to the beginnings of the universe. The development of all measurement technologies is an interdisciplinary feat that requires expertise in physics, engineering, chemistry, biology, data science, computation, and many other fields to create accurate and precise tools. We focus on the topic of measurement because it is critically important to every aspect of the chemical sciences, and advances in measurement are both driven by and help to drive chemical discovery.

There have been many reports and review articles that cover different aspects of measurement in the chemical sciences, and every area of chemical research benefits from frequent new advances in measurement. The committee chose to focus on advances in measurement and analytical chemistry that benefit all disciplines of the chemical sciences. These areas include improvements in the visualization of matter, the enhanced speed of taking and analyzing measurements, and the increased accessibility of measurement technologies. All of these improvements also play a role in making the practice of chemistry more sustainable and contribute to advancements that will help address grand challenges such as climate change.

Improvements in Visualization and Imaging

The ability to sensitively and accurately image chemical matter has increased dramatically over the past several years. To understand imaging and visualization, a Natioal Academies' report from 2006 on Visualizing Chemistry: The Progress and Promise of Advanced Chemical Imaging defined chemical imaging as "the spatial (and temporal) identification and characterization of the molecular chemical composition, structure, and dynamics of any given sample". This definition outlines the fact that chemical imaging has contributed to visualizing the very small (single atoms or molecules) to the very large (entire organs and ecosystem dynamics) by decoding the chemical and atomic composition of a sample.

Some of the most publicized improvements to measurement have been in the area of microscopy, for which Nobel Prizes were awarded in 2014 and 2017 for, respectively, the "development of super-resolved fluorescence microscopy" and "developing cryo electron microscopy for the high-resolution structure determination of biomolecules in solution". Both of these techniques enabled measurement of different types of materials, reactions, chemical compounds, and biological molecules in unprecedented ways. With cryo-electron microscopy (cryo-EM), for example, computation-enabled improvements in resolution of structural images of biomolecules improved from around 15 angstroms (Å) before 2014 to as low as 1.2 Å in 2021. Single-molecule fluorescence microscopy has not only allowed for a clearer spatial and temporal understanding of molecules but also provided an important technology for tracking reaction rates, including those driven by electrocatalysis. This technique, along with other measurements of electrocatalysis such as surface plasmon resonance, will help drive the basic chemical research needed for the optimization of fuel cells.

Increased imaging sensitivity has allowed researchers to deconvolute the chemical composition of complex samples, including in environmental research. One example is identifying and quantifying different plastic polymers from water samples. It can be quite difficult to determine which plastic is being measured, due to structural similarities in their polymers. The most common method for interpreting the composition of these samples is by pyrolysis-gas chromatography/mass spectrometry (Py-GC/MS), in which the sample is decomposed under heat and separated using gas-phase chromatography before the mass is analyzed. The full image developed from a Py-GC/MS data set helps researchers to understand the amount of different types of plastic in the water and can even lead to source identification. Similar types of MS methods have been applied to measuring the chemical interactions between different species within microbial communities in order to get a snapshot of community interactions.

In addition to optical imaging, EM, and MS, there are a number of other imaging techniques that are rapidly improving and contributing to chemistry. These include atomic and molecular spectroscopies, other types of optical imaging, nuclear magnetic resonance (NMR), noninvasive imaging, and quantum imaging. Continued improvement in our understanding of basic chemical principles will enable analytical chemistry to better utilize all of these different techniques, and those insights will continue to improve imaging technologies for chemical research, as well as research in a variety of other fields.

One area on this list that has generated a lot of interest is the use of quantum principles to enhance sensing and imaging technologies. In a recent review, researchers note that "a molecular approach to quantum sensing offers the unmatched combination of atomic structural control and tuanbility, enabling trans-formative discovery in fields spanning biology to astrophysics." The impacts that quantum physics can have on chemical discovery as well as the ways that chemistry can contribute to our understanding of quantum mechanics include a plethora of ideas and avenues, especially in a well-explored area such as quantum sensing. A National Academies' study is currently exploring this topic, specifically how quantum information science and measurement and modeling in chemistry can inform and advance one another.

Real-Time Chemical Measurements

Real-time chemical measurements and analytics have become critical for everything from monitoring the time courses of reactions to ensuring acceptable air, food, and water quality. In this report, the concept of real-time chemical measurements includes near real-time chemical measurements where a sampling phase is required in order to gather an appropriate amount of analyte to produce an accurate measurement.

The process of near-real-time chemical measurement still happens quickly and provides time-dependent and actionable information. The speed and accuracy of real time measurements regularly improve, and those improvements are reliant on advances in computation, automation, data analysis methods, engineering, and separation science. The discoveries in analytical chemistry that drive these processes create new tools and technologies that are used to identify differences in an established baseline, make experimental comparisons, and identify and intervene in dangerous or life-threatening situations.

Real-time chemical imaging and measurement can be applied to monitor changes in the environment such as monitoring volcanic activity or surveying water sources for pollutants and other unusual activity. One notable example is the Ocean Observatories Initiative, funded by the National Science Foundation (NSF), that takes data measurements on "physical, chemical, geological and biological properties and processes from the seafloor to the air-sea interface" and makes the data available to any researcher who wants to use them.

Another example is the mobile chemical analysis stations that have been deployed to identify pollution in rivers and are, in one particular case, able to take measurements of "temperature, total phosphorous, pH and ammonium ions, dissolved oxygen, conductivity, nitrate ions, and total organic carbon". These measurements rely on different types of electrochemical and optical sensors, and due to the extended duration of measurement (sometimes decades), there is a strong established baseline, meaning that abnormal levels of various pollutants are discovered instantly.

Real-time measurements of the environment can also utilize other types of instrumentation such as MS. Much research is still needed to improve sensitivity and deconvolute complex samples using rapid separation science. These types of systems are also critical for chemical manufacturing, enabling workers to monitor product quality and ensure that pollutants are not spilling into the surrounding environment.

Measurement speed is related to the instrumentation and how well analytical technologies can accurately quantify or image a subject. But it is also directly tied to the availability of rapid data analysis. High-speed data analysis requires both computing power and efficient algorithms that can quickly process a data input.

Analysis of chemical measurements has benefited greatly from the use of data interpretation algorithms that use ML and AI principles. For example, a variety of ML algorithms can be paired with electrochemical sensors to "extract complex relationships between chemical structures and their electrochemical properties and to analyze complicated electrochemical data to improve calibration and analyte classification". In these cases, ML methods help to calibrate the sensors and enhance the sensitivity of measurement, creating faster measurement capacity without the need to adjust the setup by hand. The incorporation of advanced computational techniques, such as AI and ML, into real-time measurement tools will increase the speed with which researchers and other users can make decisions based on new measurement data.

Increased Accessibility of Chemical Measurement

Many chemical measurement tools are getting smaller, more portable, and cheaper. These developments contribute to the increased accessibility of chemistry. Although not every analytical technology has benefited from increased accessibility, this is a definitive trend in a number of different areas. In the 2017 Annual Review Issue of Analytical Chemistry, it was noted that, "[c] learly, analytical tools are getting cheaper and more available as seen in reviews on paper microfluidics, 3-D printing, digital assays, and point of care diagnostics". Miniaturization of chemical instrumentation is partly enabled by the silicon revolution, and in part because chemical knowledge of sample preparation, separation, and sensor technology has facilitated the engineering of smaller devices. Frequently, decreasing the size of an instrument comes at a cost of reduced sensitivity and accuracy, but some recent advances have shown that this is not always the case.

Some instrumentation such as MS has recently undergone a transformative miniaturization process. Mass spectrometers that used to require entire rooms and specialized setups now can fit in the palm of your hand. Of course, full-room MS setups still exist and provide state-of-the art analysis of chemical samples, but there is a critical need for MS capabilities in places where entire rooms cannot be set aside for instrumentation. For example, companies such as 1st Detect Corporation and 908 Devices are building MS instruments that weigh less than 4 lbs for use in quality control of food and water as well as many security and forensics applications. The scientific community also needs analytical capabilities at places that are challenging to get to, such as the bottom of the ocean or outer space. MS technologies have already been deployed for use in space research, but their capabilities are limited and there is a need for continually miniaturizing MS systems, so that single instruments can be used for purposes such as geological, chemical, and biological testing in space.

Future of Measurement

Analytical chemistry and associated measurement technologies have made significant advances in the past couple of years, accompanied by advances in data acquisition, computation, sample preparation, microfluidics, and other enabling technologies. These tools can measure compounds in complex samples, create 3-D images of material by multiplexing different analytical methods, and trace material back to its original source with surprising accuracy. Two of the most important things that have facilitated these rapid advancements in analysis and measurement are the availability of data and the computational speed with which we can analyze data.

As measurement continues to advance, so will the complexity of the data, which will require both better ways to integrate measurement and analysis and demand for new ways to store, retrieve, and use data. One increasingly popular concept in this regard is "democratic analytical chemistry" (Figure 11.1), which relates closely to the notion of accessible chemistry.

The idea of democratic analytical chemistry is that measurements and instrumentation are available to any user directly (taking the measurements yourself) or indirectly (analyzing the measurement data from others). This concept is empowered by open access software and readily available data repositories. As more systems like this develop, the network of chemists who can use and analyze data from measurements will expand significantly.

The concept of democratic analytical chemistry is also inherently a green chemistry concept. This is because instrumentation that already exists will be used to its full potential, mitigating the need to use materials and energy to make and operate new instrumentation. Additionally, in some cases, no use of instruments is required. Rather, existing data, analyzed using the latest analysis techniques, will provide new insights.

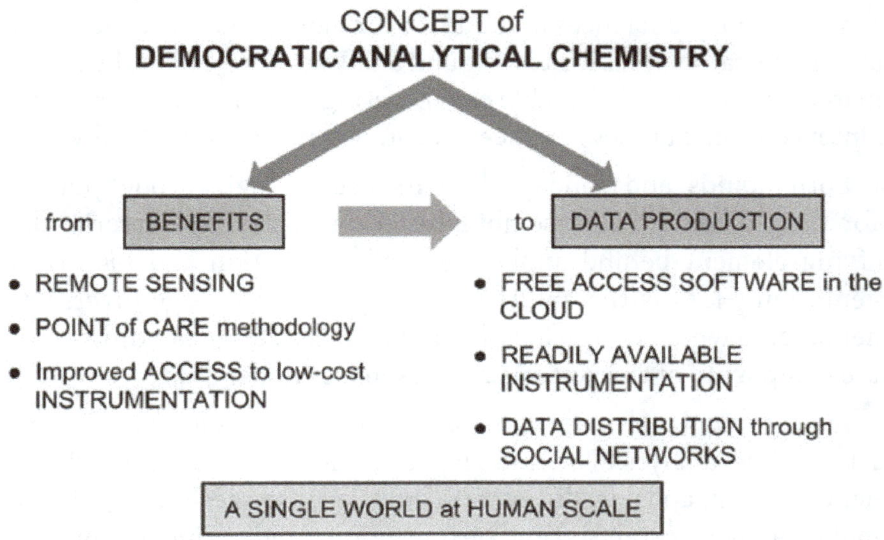

Fig. 11.1 Descriptive model of democratic analytical chemistry

Another future area for chemical measurement is the ability to perform real-time remote analytics. As discussed above, most real-time analytics rely on sample collection at a particular site and the movement of that sample to an on-site or nearby measurement device. There are, however, some unique capabilities that would be afforded by completely remote sensing. Some remote sensing techniques are already a reality, but there is much development still to pursue in this area. Real-time remote chemical sensing would afford extensive opportunities to environmental and forensic science, national security, manufacturing, and many other areas.

Elsewhere in this report, there are other examples of chemical measurement's role in advancing environmental stewardship and sustainability. As noted earlier, many new tools and technologies are improving our ability to analyze air and water quality so as to be good stewards of the environment. Beyond the more direct impacts, the idea of "green analytical chemistry" is not a new one; it evolved simultaneously with the concept of green chemistry. These ideas are mostly related to reducing waste and preventing destructive interactions with the environment. As analytical chemistry increases in complexity, there is also an opportunity to be more innovative in the implementation of green and sustainable practices in chemical measurement.

Automation

The earliest mention of laboratory automation is from 1875 when chemists developed a device to "wash filtrates unattended". Over the next century, a number of unattended processes were developed for the chemical sciences. Laboratory automation in its current form, which utilizes robotics and computation, stems from the "demands in the life sciences in the 1980s for more productive means to aliquot and dilute biological samples for testing and analysis". This led to the development of automated liquid handlers capable of generating arrays of samples with minimal manual intervention. Further advances in equipment to automatically

dose both liquids and solids led to its use in high-throughput screening assays for drug discovery and combinatorial chemistry for material discovery. The unifying element behind applications of automation and high-throughput experimentation (HTE) is the ability to rapidly generate large diverse arrays of experimental data where first principles and rational design are difficult, owing to the innate complexity of the systems and responses under study.

High-throughput techniques built upon laboratory automation technology, coupled to in-line analytical monitoring and measuring technologies, statistic experimental design, and parallel experimentation, are accelerating the discovery of new molecules (e.g., catalysts) and new reactions, and optimizing processes that generate an array of chemical products such as pharmaceuticals, agrichemicals, and materials. The capabilities of automation increased significantly in the past decade with advances in software, computational power, robotics, analytical measurements, data science, and ML now incorporated into automated experimentation.

Advances in science are made by iterative cycles of experimentation, data analysis, and hypothesis formulation. These advances progress at a rate that is commensurate with the rates at which experiments can be performed and data can be analyzed. Thus, scientists have been motivated to automate the execution and analyses of experiments to accelerate discovery. The power of automation in research is readily apparent from recent breakthroughs at the interface of biology, chemistry, physics, and engineering. Some examples are the automated synthesis and analysis of DNA, RNA, and polypeptides. Automation technologies for these biopolymers have played a critical role in most of the transformative discoveries and advances in biology and medicine. Preparation of the mRNA vaccines for SARS-CoV-2 and sequencing of the human genome are just two recent landmark achievements in synthesis and analysis, respectively, that were enabled by automation.

While chemists can certainly claim partial ownership of the invention of automated methods for the synthesis and analysis of biopolymers in biological research, scientists are now using automation in new ways to advance research in the chemical sciences. This section highlights how chemistry is benefiting from automation technologies.

Future of Automation in Chemistry

Automation is transforming the chemical sciences and manufacturing, and there are undoubtedly more advances ahead that will push against conventional boundaries. Importantly, automation, HTE, and data capture can play an important role in sustainability, especially in processes where automation can help optimize green and sustainable reaction conditions. These technologies can also help with sustainability in areas such as early-stage drug discovery where microscale or nanoscale HTE uses minute amounts of chemicals to find the desired reaction condition, and data science will enable researchers to get to the end point with far fewer experiments. Additionally, automation is enabling small-molecule syntheses on scales where computational data can be stored in the form of small molecules.

Because automation is a central part of a rapidly changing landscape in the chemical sciences, academic institutions may want to reconsider both their curricula and research avenues to minimize the gaps among basic science, translational research, and manufacturing. In this regard, Carnegie Mellon University is at the vanguard. The university has partnered with Emerald Cloud Lab to establish the Carnegie Mellon University Cloud Lab (CMU, 2021). Similar to the Molecule Maker Lab Institute, this remote-controlled lab is providing a universal platform for AI-driven experimentation in the biological and chemical sciences. It contains state-of-the-art instrumentation that can execute all aspects of daily work, from experiment design to data acquisition and analysis. The university envisions it as a resource not only for its faculty, students, and staff but for scientists around the world. It is easy to imagine that the successes of these laboratory platforms will inspire other institutions to break out of conventional ways of doing science and to embrace automation.

Computation

Building on concepts and theories from chemistry and physics developed in the late 18th, 19th, and early 20th centuries, scientists began turning theory into computable algorithms around the early 20th century, inaugurating the field that became computational chemistry. Early discoveries in computational chemistry include development of quantum chemistry and theories of chemical bonding of atoms in molecules, the chemical ensembles of statistical thermodynamics that describe interactions between molecules, and elucidation and calculations of rates of transformation in chemical reactions. Even ML has an early precursor from chemistry in the work of Louis Hammett, who elucidated a mathematical relationship between equilibrium constants and descriptors of varying substitutions on a benzene core.

Initially, solutions to these algorithms were calculated manually, using numerical methods for integration, differentiation, enumeration of ensembles of states, and curve fitting, with the result that algorithms were applied only to small, more tractable systems such as the hydrogen atom. Coming up with these solutions could take thousands of person-hours to calculate. The advent of mechanical, electromechanical, and vacuum based computers vastly increased the speed of calculations and concomitantly expanded the size and complexity of the systems that could be studied. ENIAC, or Electronic Numerical Integrator and Computer, the first programmable, general-purpose computer, could perform thousands of computations per second.

Hand-in-hand with this computational speed, ENIAC's architecture of branching and stored memory enabled programmers to move beyond a linear sequence of operations and incorporate conditional execution (if-then-else for loops) in their codes. This innovation in computer architecture led directly to computational innovation: the first-ever computational Monte Carlo simulation, a calculation run on ENIAC on behalf of Los Alamos National Laboratory that simulated the movement of neutrons through the fissile material of a nuclear bomb (Haigh et al., 2014). Since this first calculation, Monte Carlo simulations have become a widespread and valuable tool across a wide range of computational chemistry applications. This synergy among computational speed, computer architecture, and algorithmic innovation is a recurring theme throughout the development of computational chemistry as a discipline.

There have been many reports and review articles that cover the current state of the art for computational chemistry, and every area of chemical research benefits from frequent new advances in computational power and algorithm development. While this chapter and many other sections in the report point out some specific and important examples, the committee chose to focus on chemistry-enabled advances in computational hardware that benefit all disciplines of the chemical sciences and that have the potential for transformative impact on the types of chemistry challenges that can be addressed by computational chemistry algorithms.

Future of Computational Chemistry

Miniaturization, increased computational power, and faster computational speeds have vastly increased the size and complexity of chemistry problems accessible to computation, but we are reaching the limits of miniaturization. As transistors are reaching scales at the size of atoms, we can go no smaller. In addition, as computation has become a regular and important component of chemical research, and data science becomes embedded across a wide range of disciplines, energy consumption becomes a limiting factor.

Developments for the future will need to focus on improving sustainability and energy efficiency across all scales of computation—supercomputers, data centers, and personal computing devices—and on new computing paradigms that allow us to move beyond Moore's Law. As we explore new materials and computer architectures, chemistry and chemical engineering will continue to play a role in making fundamental discoveries to drive sustainability in computation, just as they drove miniaturization in the past.

As was the case with development of our current semiconductor architectures, new materials and computer architectures will drive algorithmic innovation that enables us to ask new chemistry questions in entirely new ways.

Quantum Simulation with Quantum Computers for Chemistry

Nonclassical computing architectures such as quantum computers represent a new class of computational tools for chemistry simulations. These computer architectures will complement classical tools since performing certain calculations with quantum computers provides a computational resource advantage for specific problems in quantum chemistry. Quantum computers behave fundamentally differently from classical computers because they are made of different types of materials than semiconductors and store and encode information based on the principles of quantum mechanics.

Though the development of quantum computers is an active area of research, several classes of them have begun to emerge. At a high level, these include ion trapping, transmon-based technologies involving superconductors, and theoretically topological quantum computers. Quantum bits, "qubits," are being produced today that leverage ion-trapping materials and superconductors. (At the time of this report, topological quantum computers had not been actualized.)

Since quantum computers operate under fundamentally different principles than do classical computers, they are best suited to tackle certain types of problems. As a concrete example of the differences in problem solving, there is no advantage (speed up) to using a quantum computer for addition or multiplication operations, but there is a theoretical advantage for number factoring (e.g., Shor's algorithm).

Through the construction of quantum circuits, programmers leverage quantum mechanical properties of the qubits—such as superposition, entanglement, and interference—for computation. Thus, quantum computers are best suited to handle complex, interconnected information.

Chemists may intuitively associate complex, interconnected system problems with electron correlations, and indeed, the behavior of electrons in a molecular system is well described by quantum algorithms. Quantum chemistry calculations using Schrödinger's equation are typically exponential in resource requirements. As the problem size gets larger, the calculation quickly becomes intractable on classical devices. So, "exact" calculations on molecules, where the entire set of particle interactions in a system is simulated, become impossible.

Furthermore, the Pauli exclusion principle adds to the cost because two electrons cannot sit at the same energy with the same spin, which is referred to as the "sign problem." State-of-the-art full-configuration interaction calculations are intractable past approximately 20 orbitals and approximately 20 electrons, even with today's most powerful supercomputers.

The calculations can take more than a week to run, and calculations for reactions with popular approximate computational chemistry methods such as density functional theory (DFT) can be inaccurate. To mitigate these problems, simulations involving DFT invoke approximations to avoid high computational cost––for example, modeling the distinct electrons in an "electron field" compared to calculating all of the distinct electron– electron interactions. Determining the energy landscape for a reaction mechanism can take months or years of trial and error.

Although all of these issues with classical computers might be solved using quantum computers, there is still a large overhead associated with quantum chemistry calculations on quantum computers. A quantum chemistry circuit can require millions of gates to do a full configuration interaction calculation of a molecule. For quantum algorithms, such as quantum phase estimation for chemistry simulations, error-corrected devices (e.g., universal fault-tolerant quantum computers) will be necessary in order to carry out the operations.

Quantum computing in a fault-tolerant regime promises more accurate chemistry calculations (possibly leading to predictive chemistry). However, while research continues in quantum hardware, software, and theory, researchers can test small examples on devices available today. Researchers will often use a simulator that runs on a classical computer to test resources for calculations before trying the calculation on a noisy real device. The quantum computers all have differences in connectivity, number of qubits, and noise profiles, which enables customization for a calculation of interest. How to best leverage the hardware for chemistry calculations remains an open research question.

Beyond quantum chemistry, quantum computing might have advantages for chemistry broadly, through quantum ML, optimization, and Monte Carlo calculations. Applied to chemistry examples, it is possible to imagine exploring quantum computers for classification, anomaly detection, conformation optimization, solvation, and kinetics, to name a few. Finally, it is feasible that due to hardware performance differences, different types of quantum computers may be better suited for different classes of problem solving. Thus, incorporating quantum computers into computational chemistry workflows is a promising use of this emerging technology.

Catalysis

Catalysts are used in most of the chemical reactions that make commonplace products, such as fuels, food, pharmaceuticals, synthetic fibers, and plastics. They drive chemical processes by facilitating the rearrangement of atoms and molecules. It is estimated that more than 95% (by volume) of everyday products contain ingredients that are made using catalysts, accounting for >80% of added value in the chemical industry. Despite the well-established importance of catalysis, feedstock changes, energy transitions, and other environmental, health, and safety concerns will bring about fundamental changes in the chemical economy requiring new approaches for catalytic systems as well as a reexamination of previously explored technologies. In particular, the fields of electrocatalysis, photocatalysis, and biocatalysis are expected to play significant roles in the resource-efficient conversion of emerging feedstocks.

Reemergence of Photocatalysis, Electrocatalysis, and Biocatalysis

While alternative methods to classical metal-based catalysis, such as photocatalysis, electrocatalysis, and biocatalysis, have been under consideration for decades as useful methods to drive chemical reactions, they are recently experiencing a resurgence in popularity as alternative methods to classical metal-based catalysis. The literature describing new transformations via electrochemistry, photochemistry, and biocatalysis portends their potential for vast applications in materials synthesis, chemical production, and new pharmaceutical and agrochemical molecules. Though there are many challenges to be overcome, these methods provide an alternative to energy-intensive synthesis methods that use high heat, high pressure, or both to perform chemical transformations in the production of some industrial chemicals.

Future Challenges in Catalysis

Fundamental challenges in catalysis and surface science are articulated in two DOE reports. To address these challenges, DOE established five research priorities: considering both the binding site and allosteric effects when designing catalysts, understanding the dynamic evolution of catalysts, manipulating reaction networks in complex environments to selectively steer catalytic transformations, designing electrocatalysts that are highly selective and energy efficient, and driving new catalyst discoveries by coupling data science, theory, and experiment. Research programs and centers funded by the DOE Office of Basic Energy Sciences are addressing several of these challenges.

The goal is that fundamental science advances and tools made available by this research will lead to precision design of catalysts that maximize yields of desired products under mild conditions, thus conserving both feedstocks and energy. While catalysis has been instrumental in major industrial success stories such as ammonia synthesis and the catalytic converter, there are processes that have eluded a catalytic solution for decades. An example is the steam cracking of naphtha to produce ethylene, a major platform chemical with an annual capacity of 160 million tons. Among petrochemical processes, this one consumes the largest amount of energy with the endothermic cracking reaction accounting for roughly 13–22% of overall energy and the downstream separation of ethylene from ethane accounting for the rest.

Ongoing efforts to power the steam cracking furnaces with renewable electricity (the so-called electrification of steam crackers) are expected to significantly reduce the carbon footprint of this energy-intensive process. Efforts to develop a catalyst that further reduces energy requirements for this endothermic process have been challenged by rapid catalyst deactivation due to coking and catalyst stability issues. Advances in heterogeneous catalysis, materials science, and process engineering are needed to effectively decarbonize current ethylene manufacturing.

Many industrial heterogeneous catalytic reactions, including the most relevant reactions related to energy and decarbonization, are limited by the Sabatier principle. The activity of a catalytic site is dictated by one or more rate-determining steps in the catalytic sequence of adsorption, surface reaction, and desorption. Overcoming this limitation is a fundamental research challenge in heterogeneous catalysis. Recently, researchers put forward the catalytic resonance theory to understand and potentially enhance rate and selectivity in heterogeneous catalysis beyond the Sabatier limit.

The theory recognizes that surface binding energy and transition state energies oscillate over time such that at a certain point, it may be easier for a reactant to adsorb on, or a product to desorb from, the catalyst surface. By superimposing an external wave at the catalyst surface that resonates with these oscillations, the simulation work shows that it may be possible to accelerate a surface-catalyzed reaction by several orders of magnitude compared to static conditions. The resonance theory was demonstrated experimentally for intensified formic acid formation via dynamic electrocatalysis.

However, creating the required frequency of temperature or concentration oscillations in thermally driven conversions may be more challenging. This is an example of fundamental chemical research where experimentation and theory intersect to produce some groundbreaking possibilities for catalytic processes.

Plasma-driven heterogeneous catalysis is another area that shows promise for the chemical transformation of hard-to-activate molecules such as N2, CO2, and CH4, all relevant in ammonia synthesis and in mitigating greenhouse gas emissions. Researchers note that there is a challenge in discovering catalytic materials that can take full advantage of a given plasma operating environment to selectively produce desired products. Key to the discovery of such materials is the development of relationships between plasma configurations, catalytic materials, and plasma-generated species by carefully and controllably combining plasmas with catalysts and isolating contributions of the catalyst from those of the plasma.

For example, using ozone, a powerful oxidant, at mild conditions is being tried in organic syntheses in flow reactors and light alkane activation. Electrification of the chemical industry using affordable renewable energy could spur increased use of plasmas and ozone in chemical transformation schemes.

Multiphase catalysis, involving gas, liquid, and solid phases, may be key to profitably processing emerging feedstocks such as biomass, CO2, recycled plastics, and natural gas liquids. For example, catalytic hydroprocessing of biomass to produce fuel and chemical precursors is often performed in a liquid phase. There is growing interest in performing catalytic CO2 hydrogenation in the liquid phase to produce fuels and chemicals. Similarly, chemical upcycling of plastic waste involves catalytic hydroprocessing in a melt phase.

To understand how the multiphase environment around a catalytic site influences its elementary reaction steps, advanced experimental operando/in situ tools are needed to probe catalytic surfaces in condensed media under pressurized environments. Recent examples include the development of high-pressure operando x-ray absorption spectroscopy and NMR techniques for probing heterogeneous catalytic reaction mechanisms involving conversion of biomass model compounds.

The broad field of catalysis, encompassing heterogeneous, homogeneous, bio-, electro-, and photocatalysis, will be key to solving many of the challenges related to climate change and ensuring a sustainable supply of energy and materials. Success will require building a bridge between theory and experiments to understand how intrinsic material properties determine catalyst performance: what some people call the quest for the "catalyst genome". This will require fundamental chemical research that integrates expertise in materials, measurement, and computation.

1. Research on catalysis covers broad length and time scales. Ideal catalytic conditions on a small scale may not produce the desired product at larger scale. To bridge this "scale gap," it is vital to consider reaction and process engineering aspects along with life-cycle analysis during early stages of catalyst and process development. This will ensure that energy efficiency is considered across all manufacturing steps, including separation processes.

The urgent need to tackle sustainability-related challenges is driving the need to improve the pace and efficiency of discovering new catalysts and developing new catalytic processes. Toward this end, available materials, adsorption, and reaction data are being harnessed using high-throughput computation and ML to gather further insights and predict new materials.

However, this knowledge extraction approach is challenged by a data deficit, or non uniform reporting of materials-related data that are not computer readable. Researchers note that "in order to make data widely useful, rather advanced and well-coordinated approaches are needed that are beyond what a single group or institution can develop and sustain." Digitalization of the catalysis field is essential to "enable efficient data driven interdisciplinary development of catalysts and catalytic processes". The NFDI4Cat (National Research Data Infrastructure for Catalysis-Related Sciences), supported by the German government, is an example of such a large-scale effort.10 Establishment of an "internet of catalysis" will guide research along the development chain from molecules to chemical processes. The creation of digital workflows that bridge theory and experimental studies in catalyst design, characterization, kinetics, and related engineering aspects will accelerate discovery and innovation in the catalysis sciences.

CONCLUSION

RECAPITULATION OF KEY CONCEPTS

In a chemical reaction, both the molecules and atoms, which interact with each other, are known as reactants. In a chemical reaction, both the molecules and atoms produced by the reaction are referred to as products. Only the atoms accessible in the reactants can end up in the products in a chemical reaction. No new atoms are either created or destroyed. In a chemical reaction, reactants usually contact each other, bonds between atoms in the reactants are broken, and the atoms rearrange and form new bonds to form the products. In a chemical change, a new compound can be formed. Whereas, in a physical change, the substance changes its existence state. Ions or atoms, or molecules that react to form a new substance, are known as reactants; the new molecules or atoms formed are called products. A chemical reaction is the one that always follows the law of conservation of mass. It means, no atom is either created or destroyed, but only a new product is formed from reactants.

Advances in measurement, automation, computation, and catalysis have already driven chemical research forward, and there are many examples where advances in these tools have created subsequent advances in chemistry. Though there are a large number of enabling technologies that move chemistry forward, the committee identified these four because they are considered to have the biggest immediate impact and offer enormous potential for the future. The committee also noted some similar needs in these four emerging areas, especially around the use of chemical data. The committee reached a number of conclusions about the importance of measurement, automation, computation, and catalysis for enhancing the chemical economy.

- Chemistry is an enabling scientific discipline that will continue to have the largest impact on society when chemists collaborate with experts from other areas such as engineering, biology, physics, computation, and data science to generate new fundamental knowledge and create translational impact at larger scales.

- Measurement, automation, computation, and catalysis are the enabling tools and technologies of fundamental chemical research that will have a substantial impact on both the adoption of novel methodologies and future discoveries in the chemical economy.
- The ability to collect, document, store, share, and use chemistry-related data is needed to advance the use of new tools, such as computation and automation in fundamental chemical research, and increase the accessibility of chemical research to a larger community of practitioners. This information architecture will produce an indispensable tool for the chemical sciences research community to increase the pace and efficiency of innovation by fully harnessing advances made with previous research investments. In considering each of the four areas outlined in this chapter, important conclusions emerged related to measurement, automation, computation, and catalysis.
- Analytical chemistry will continue to play a substantial role in driving fundamental chemistry and having an important impact on numerous sectors including medicine, environmental and forensic monitoring, and national security, especially as analytical tools increase in speed and accuracy and instrumentation miniaturizes.
- Automation has the potential to change the way chemistry research is designed by increasing the number of syntheses, measurements, or process steps that can be done in rapid succession and by producing data that can be analyzed to help move research forward.
- Fundamental research in computational chemistry is fundamental research in chemistry. A synergistic relationship among advances in computer architectures, computational chemistry algorithms, and the application of computational chemistry enables innovation in all chemistry disciplines.
- Fundamental multidisciplinary research in chemistry, physics, and engineering has played a critical role in the ongoing development of modern computer architecture, and will continue to do so with the continued miniaturization of computers and the emergence of new architectures such as quantum and neuromorphic computing.
- Advances in catalysis remain critically important for driving new products and processes, and the subfields of catalysis covering heterogeneous, homogeneous, biocatalysis, electrocatalysis, and photocatalysis will all play key roles in advancing the fundamental science and technologies needed for making renewable energy, decarbonizing the chemical industry, and promoting a circular economy.

BASIC CONCEPTS OF CHEMICAL REACTIONS

- A Chemical Reaction is defined as a process that takes place when either two or more molecules interact to produce a new product (s).
- Compounds, which interact to produce new compounds, are known as reactants, whereas the newly formed compounds are known as products.
- Chemical reactions play an integral role in various industry verticals, customs, and even in our daily lives. They are continuously happening in our surroundings; for example, pottery, rusting of iron, fermentation of wine and more.
- In a chemical reaction, a chemical change should take place, which is generally noticed with physical changes like heat production, precipitation, color change and so on.
- A reaction can occur between either two ions or atoms or molecules, and they form a new bond, and no atom is either created or destroyed, but a new product is formed from reactants.
- The rate of reaction completely depends on and is affected by factors like temperature, pressure, the concentration of reactants.

Chemical Equations

Due to the excess amounts of chemical reactions happening around us, a nomenclature was developed to simplify how we express the chemical reaction in a chemical equation form. A chemical equation is nothing but a mathematical statement that symbolizes the product formation from reactants while stating certain conditions for which the reaction has been conducted.

The reactants lie on the left-hand side, whereas the products formed on the right-hand side are connected either by one-headed or two-headed arrows. For example, a reaction is given below:

$$A + B \rightarrow C + D$$

Here, both A and B are the reactants that react to form the products C and D. In an actual chemical equation; reactants can be denoted with the help of their chemical formula. To assure the law of conservation of mass, a chemical equation should be balanced. The number of atoms present on both sides must be equal. This is said to be equation balancing.

Let us consider an actual chemical reaction between Oxygen (O_2) and Methane(CH_4), which can be shown as:

Here we can observe how the number of every atom on the left side is balanced on the right side, which is stated by the law of conservation of mass.

Types of Chemical Reactions

Let us discuss the types of chemical reactions in detail with examples.

The basis for various types of reactions is given as the product formed, the changes that take place, the reactants involved, and so on. Some types of reactions are given below:

- Combustion reaction
- Decomposition reaction
- Neutralization reaction
- Redox Reaction
- Precipitation or Double-Displacement Reaction
- Synthesis reaction

Combustion Reaction

A combustion reaction is defined as a reaction having a combustible material with an oxidizer to form an oxidized product. An oxidizer is given as a chemical fuel that requires burning, generally oxygen. Consider the example of the combustion of magnesium metal as given below.

$$2Mg + O_2 \rightarrow 2MgO + Heat$$

Here, 2 magnesium atoms get to react with a molecule of oxygen-producing 2 molecules of the compound magnesium oxide-releasing some heat in the process.

Decomposition Reaction

A Decomposition reaction is defined as a reaction where a single component breaks down into multiple products. Certain energy changes in the environment have to be made like light, heat, or electricity breaking bonds of the compound. Consider an example of the decomposition of calcium carbonate forming CaO (Quick Lime), which is a primary component of cement, as given below:

$$CaCO_{3(s)} \rightarrow Heat \rightarrow 5Ø;Ü5ØRÜ5ØNÜ5ØaÜ\ CaO_{(s)} + CO_{2(g)}$$

CaCO3 (s) CaO (s)+CO2 (g)

Here, the compound Calcium carbonate, when heated, breaks down into Carbon Dioxide and Calcium Oxide.

Neutralization Reaction

A Neutralization reaction is basically defined as the reaction between an acid and a base by forming salt and water as the products. The water molecule that is formed is by the combination of H+ ions and OH– ions. The entire pH of the products when a strong acid and base undergo a neutralization reaction will be

7. Consider an example of the neutralization reaction between Sodium Hydroxide and Hydrochloric acid giving out sodium chloride (called Common Salt) and water.

$$HCl + NaOH \rightarrow NaCl + H_2O$$

Here, the Hydrochloric acid and Sodium Hydroxide (an acid and a base) react in a neutralization reaction to form Sodium Chloride (called Common Salt) and water as the products.

Redox Reaction

A REDuction-OXidation reaction is defined as a reaction where there is a transfer of electrons between the chemical species. Let us consider an example of an electrochemical cell-like redox reaction between Hydrogen and Zinc, which is chemically shown as follows:

$$Zn + 2H+ \rightarrow Zn^{2+} + H_2$$

Here, a Zinc atom reacts with 2 ions of the positively charged hydrogen, where the electrons get transferred from the zinc atom and hydrogen becomes a stable molecule and Zinc ion is the resultant product.

Double-Displacement or Precipitation Reaction

This is a type of displacement reaction (which is the precipitation reaction), where two compounds react, and consequently, their cations and anions switch places, forming two new products. Consider an example of the reaction between sodium chloride and silver nitrate. The products will be sodium nitrate and silver chloride after the double-displacement reaction, which is shown below:

$$AgNO_3 + NaCl \rightarrow AgCl + NaNO_3$$

Here, Sodium Chloride and Silver Nitrate undergo a double displacement reaction. Wherein Silver replaces the Sodium in Sodium Chloride and Sodium joins with Nitrate becoming Sodium Nitrate along with the Silver Chloride as the resultant product.

Synthesis Reaction

A Synthesis reaction is the most basic type of reaction wherein various simple compounds combine under certain physical conditions resulting in a complex product. Here, always, the product will be a compound. Let us consider the Synthesis reaction of sodium chloride with reactants chloride gas and solid sodium, as shown below:

$$2Na_{(s)} + Cl_{(g)} \rightarrow 2NaCl_{(s)}$$

Here, there are 2 Atoms of solid Sodium reacting with Chlorine gas forming in Sodium Chloride viz. Common Salt as the product.

Reactivity is a chemistry term that describes how easily a substance can conduct a chemical reaction. The substance can react individually or with other atoms or compounds, and the reaction is usually accompanied by the release of energy.

The most reactive elements and compounds have the potential to spontaneously ignite or explode. They usually burn in water and in the presence of oxygen in the air. The temperature has an impact on reactivity. Temperature raises the amount of energy available for a chemical reaction, thus making it more likely to react.

Chemical reactivity

In chemistry, a material's reactivity measures how quickly it triggers a chemical reaction. The substance may participate in the reaction alone or in conjunction with other atoms or compounds and an energy release generally follows it.

- The term reactivity can also refer to how rapidly a material reacts or the rate at which a chemical reaction occurs.
- The most reactive elements and compounds can spontaneously or explosively ignite.
- They usually burn in the presence of both water and oxygen in the air.
- The temperature affects reactivity. Increasing the temperature provides more energy for a chemical reaction, which improves its chance of reacting.
- The physical properties of a sample, its chemical purity, and the presence of other compounds all influence its reactivity.
- Baking soda and water are not particularly reactive;
- Baking soda and vinegar combine to form carbon dioxide gas and sodium acetate.
- The structure of the periodic table allows for reactivity predictions. Both extremely electropositive and electronegative materials are highly reactive.
- These elements may be found in the periodic table's upper right and bottom left corners.
- The size of the particles has an impact on their reactivity. A pile of corn starch, for example, is relatively inert. It's tough to start a combustion process with a direct flame on the starch. In contrast, corn starch rapidly ignites when evaporated into a cloud of particles.

Concepts of chemical reactions

- Chemical reactions solely affect the locations of electrons while forming and breaking chemical bonds between atoms, without any change to the nuclei, and may be represented using a chemical equation.

- New substances are produced during a chemical reaction. Here the chemical bonds break, and the atoms that make up the substances split and reorganise themselves into new compounds with new chemical bonds. This is referred to as a chemical reaction.
- Atoms are never formed or destroyed during chemical processes. The same atoms that exist in the reactants are also found in the products; they are simply rearranged in different sequences.

Parts of chemical reactions

- The components that interact in a chemical reaction are known as reactants.
- Products are the compounds generated as a result of the reaction.
- Chemists use a symbolic description known as a chemical equation to express the reactants and products.
- The reactants are listed on the left side, the products are listed on the right side.
- The reactants and products are separated by an arrow showing the direction of the reaction.
- While many chemical equations portray reactants yielding products, the chemical reaction usually occurs in the other direction.
- In a chemical reaction or a chemical equation, no new atoms are formed or lost (mass conservation), but chemical bonds between individual atoms can be broken and then be further established.

Balanced and unbalanced

Chemical equations can be either unbalanced or balanced. Although an unbalanced chemical equation does not account for mass conservation

It is typically a good starting point since it specifies the products and reactants and the direction of the chemical reaction.

Conclusion

During a chemical process, new compounds are created. The chemical bonds of the substances are broken, and the atoms that comprise the substances divide and rearrange themselves into new substances with new chemical bonds. When this happens, it is referred to as a chemical reaction. A chemical reaction is a process that converts one or more compounds into new compounds.

FINAL THOUGHTS ON THE DYNAMIC NATURE OF CHEMICAL REACTIVITY

Chemical equilibrium is the state of a system in which the reactant and product concentrations do not change over time and the system's attributes do not change further.

Reactions take place in both forward and reverse directions. When the rates of the forward and reverse reactions are similar in such processes, the concentrations of the reactants and products stay constant. The reaction is said to be in chemical equilibrium at this point. This equilibrium, on the other hand, is thought to be dynamic in nature. This is due to the fact that it comprises a forward reaction in which the reactants react to produce products and a reverse reaction in which the products can react to produce the original reactants.

Dynamic Nature of Chemical Processes in Equilibrium

Take a look at the following reversible reaction–

$$A + B \Rightarrow C + D$$

The products (C and D) build with time, whereas the reactants (A and B) diminish. As a result, the rate of forwarding reaction decreases while the rate of reverse reaction increases. Finally, both reactions take place at the same moment, resulting in a state of equilibrium. This balance can be attained from either direction.

Haber's Process

Fritz Haber, a German chemist, invented a method for producing ammonia from dinitrogen and dihydrogen. This is known as the Haber process.

$$N_2(g) + 3H_2(g) \Rightarrow 2NH_3(g)$$

Haber began with known proportions of dinitrogen and dihydrogen, kept the mixture at high temperature and pressure, and measured the amount of ammonia generated at regular intervals. As the reaction progressed, he saw that the composition of the mixture remained constant, despite the presence of some reactants. This indicates that the reaction has achieved equilibrium.

In addition, Haber's method exemplifies the dynamic character of chemical equilibrium in the following way. Maintaining the same experimental settings as before, hydrogen was substituted with deuterium (D_2). As a result, instead of NH_3, ND_3 is produced. Both reactions, one using H_2 and the other involving D_2, were allowed to reach equilibrium. When these two mixtures were mixed and left for a while, the concentration of ammonia was found to be the same as previously. Despite this, mass spectrometry confirmed that, in addition to ammonia, all types of deuterium and dihydrogen were present.

Because the forward and reverse processes continue, the scrambling of H and D atoms must be conceivable. If the reaction had ended when it reached equilibrium, there would have been no mixing of isotopes. As a result, chemical reactions attain a state of dynamic equilibrium in which the rates of forwarding and reverse reactions are equal and no net change in composition occurs.

Chemical Equilibrium is Bidirectional

Whether we begin a reaction with reactants or products, equilibrium can be reached on both sides. Take a look at the following reaction.

If we start the reaction with equal initial concentrations of H2 and I2, the reaction will go forward, with the concentrations of hydrogen and iodine decreasing and the concentration of hydrogen iodide increasing until it achieves equilibrium. If we reverse the above reaction, the concentration of hydrogen iodide falls while the concentrations of hydrogen and iodine grow until equilibrium is reached. As a result, if the total number of atoms of an element in a given volume is the same, we get the same equilibrium mixture whether we start with reactants or products.

Conditions for Equilibrium

The process of achieving chemical equilibrium is a dynamic one. Even after the equilibrium state is reached, the forward and reverse processes continue to occur. However, the speeds of the reactions are the same in this case, and there is no change in the relative concentrations of reactants and products for an equilibrium reaction. The following are the criteria and properties of an equilibrium system.

- The system must be closed, which means that no substances may enter or exit it.
- Equilibrium is a living, breathing thing. Even if we cannot see the reactions, both forward and reverse reactions are occurring.
- The rates of forwarding and backward reactions must be equal.
- The number of reactants and products does not have to be the same. However, once equilibrium is reached, the amounts of reactants and products remain constant.

Examples of Chemical Equilibrium

- The forward reaction in chemical reactions converts reactants into products, whereas the backward reaction converts products back into reactants. There are two states, reactants and products, and they both exist in different compositions.
- When the reaction begins, the rates of the forward and backward responses may become equal after some time.

- Following this, the number of reactants converted will be created again by the reverse reaction, resulting in no change in the concentration of reactants and products. As a result, the reactants and products will be in a state of chemical equilibrium.

BIBLIOGRAPHY

Allouche R., Haoui R., Renane R., "Numerical simulation of reactive flow in non-equilibrium behind a strong shock wave during re-entry into earth's atmosphere", Mechanics & Industry, vol. 15, no. 1, pp. 81–87, 2014.

Andzelm, Jan, Jon Baker, Andrew Scheiner, and Michael Wrinn. "A density functional study of chemical reactions." International Journal of Quantum Chemistry 56, no. 6 (December 15, 1995): 733–46.

Atzler F., Demoulin F.X., Lawes M., et al., "Burning rates and flame oscillations in globally homogeneous two-phase mixtures (flame speed oscillations in droplet cloud flames)", Combustion Science and Technology, vol. 178, no. 12, pp. 2177–2198, 2006.

Babinsky E., Sojska P.E., "Modeling drop size distributions", Progress in Energy and Combustion Science, vol. 28, no. 10, pp. 303–329, 2002.

Barrere M., Prud'Homme R., Equations fondamentales de l'Aérothermochimie, Masson et Cie, Paris, 1973.

Beau P.-A., Modélisation de l'atomisation d'un jet liquide. Application aux sprays Diesel, Doctoral Thesis, CoRIA-University of Rouen, 19 December 2006.

Benjamin T.B., Ursell F., "The stability of the plane free surface of a liquid in vertical periodic motion", Proceedings of the Royal Society of London, vol. 255, pp. 505–515, 1954.

Blokkeel G., Mura A., Demoulin F.X., et al., " A continuous modeling approach for describing the atomization process from inside the injector to the final spray", ICLASS, Sorrento, Italy, 2003.

Bockhorn H., " A short introduction", Soot Formation in Combustion – Mechanisms and Models, Springer, Berlin, pp. 3–8, 1994.

Borghi R., Lacas F., " Modeling of liquid propellant spray combustion in rocket engine combustor", Second International Symposium on Liquid Rocket Propulsion, ONERA-Châtillon, 19–21 June 1995.

Boukra M., Cartellier A., Ducasse E., et al., "Use of Faraday instabilities to enhance fuel pulverisation in air-blast Atomisers", Comptes Rendus Mécanique, vol. 337, pp. 492–503, 2009.

Bozier O., Contribution à l'étude des caractéristiques de combustion isochore d'une suspension de particules solides réactives. Génération de la suspension; influence de l'état initial du mélange, Doctoral Thesis, ENSMA-Poitiers, 14 September 2004.

Bozier O., Veyssière B., "Study of the mechanisms of dust suspension generation in a closed vessel under microgravity conditions", Microgravity Science and Technology, vol. 22, pp. 233–248, 2010.

Bremond N., Clanet C., Villermaux E., "Atomization of undulating liquid sheets", Journal of Fluid Mechanics, vol. 585, pp. 421–456, 2007.

Brown P.P., Lawler D.F., "Sphere drag and settling velocity revisited", Journal of Environmental Engineering, vol. 129, pp. 222–231, 2003.

Carlson D.J., Hoglund R.F., "Particle drag and heat transfer in rocket nozzles", AIAA Journal, vol. 2, no. 11, pp. 1980–1984, 1964.

Chahine G., "Bubble Dynamics", Research and Development in Applied Sciences. Available at: http://www.dynaflow-inc.com/Publications/FullTextReports/Fulltext_Bubble_Cav.htm, 2013.

Chapman, Ricky A. "The chemical reactions of lithium monoxide with transition metals." Thesis, University of Nottingham, 1990.

Chen Y.-J., Steen P.H., "Dynamics of inviscid capillary breakup: collapse and pinchoff of a film bridge", Journal of Fluid Mechanics, vol. 341, 1997.

Chen, Kunfeng, Shuyan Song, and Dongfeng Xue. "Hopper-like framework growth evolution in a cubic system: a case study of Cu2O." Journal of Applied Crystallography 46, no. 6 (October 26, 2013): 1603–9.

Chirkina, E. A., L. B. Krivdin, A. G. Mal'kina, and B. A. Trofimov. "Quantum chemical study of mechanisms of organic reactions." Russian Chemical Bulletin 64, no. 3 (March 2015): 511–17.

Clift R., Grace J.R., Weber M.E., Bubbles, Drops and Particles, Academic, New York, pp. 306–314, 1978.

Coelho P.J., "Numerical simulation of the interaction between turbulence and radiation in reactive flows", Progress in Energy and Combustion Science, vol. 33, pp. 311–383, 2007.

Corsin S., Lumley J., "On the equation of motion for a particle in a turbulent fluid", Applied Science Research A, vol. 6, pp. 114–116, 1956.

Cousin J., Dumouchel C. "Effect of viscosity on the linear instability of a flat liquid sheet", Atomisation and Spray, vol. 6, pp. 563–576, 1996.

Cousin J., Prédiction des distributions granulométriques par le formalisme d'entropie maximum. Application à plusieurs injecteurs mécaniques, Doctoral Thesis, University of Rouen, 28 October 1996.

Criado C., Galan-Montenegro P., Velasquez P., et al., "Diffusion with general conditions in electrochemical systems", Journal of Electroanalytical Chemistry, vol. 488, pp. 59–63, 2000.

Cryer C., Steen P.H., "Collapse of the soap-film bridge: Quasistatic description", Journal of Colloid Interface Science, vol. 154, pp. 276, 1992.

Czapik, Przemys³aw, and Zdzis³awa Owsiak. "Chemical corrosion of external stairs – case study." MATEC Web of Conferences 163 (2018): 05003.

Da Silva M.L., Marraffa L., " Radiative transfer in entry flows", 10th International Planetary Probe Workshop, EDL Short Course, San José, California, United States, 15–21 June 2013.

Da Silva M.L., Simulation des propriétés radiatives du plasma entourant un véhicule traversant une atmosphère planétaire à vitesse hypersonique – application à la planète Mars, Doctoral Thesis, University of Orléans, 2004.

De Groot S.R., Mazur P., Non Equilibrium Thermodynamics, North-Holland Publishing Company, Amsterdam, 1963.

De Groot S.R., Mazur P., Non Equilibrium Thermodynamics, North-Holland Publishing Company, Amsterdam, 1969.

Delhaye J.-M., ZUBER N., Multiphase Science and Technology, Hemisphere Publishing Corporation, 1999.

Demarco Bull R.A., Modeling thermal radiation and soot formation in buoyant diffusion flames, Doctoral Thesis, University of Aix-Marseille, 2012.

Demoulin F.-X., Contribution à la modélisation de la combustion turbulente des milieux à deux phases, Doctoral Thesis, CORIA, University of Rouen, 2006.

Deshmukh K.V., Modest M.F., Haworth D.C., "Direct numerical simulation of turbulence-radiation interactions in a statistically one-dimensional nonpremixed system", Journal of Quantitative Spectroscopy and Radiative Transfer, vol. 109, pp. 2391–2400, 2008.

Dimotakis P.E., "Two-dimensional shear layer entrainment", AIAA Journal, vol. 24, no. 11, pp. 1791–1796, 1986.

Dodemand E., Prud'Homme R., Kuentzmann P., "Influence of unsteady forces acting on a particle in a suspension. Application to the sound propagation", International Journal of Multiphase Flow, vol. 21, no. 1, pp. 27–51, 1995.

Dombrowski N., " On possibility of determination of disperse composition of two-phase flow using light scattering under small angles", Teplofizika Vysokich Temperatur, 1982.

Dombrowski N., Fraser R.P., "A photographic investigation into the disintegration of liquid sheets", Philosophical Transactions of the Royal Society of London A, vol. 247, pp. 101–130, 1954.

Dombrowski N., Hooper P. C., "The effect of ambient density on drop formation in sprays", Chemical Engineering Science, vol. 17, pp. 291–305, 1962.

Dombrowski N., Johns W.R., "The aerodynamic instability and disintegration of viscous liquid sheets", Chemical Engineering Science, vol. 18, pp. 203–214, 1963.

Dordain J.-J., Prud'Homme R., Bouttes J.F., Fluides hétérogènes, Majeure Sciences de l'espace, Ecole polytechnique, 1988–1989.

Dorey L.-H., Modélisation des phénomènes couplés combustion – formation des suies – transferts radiatifs dans les chambres de combustion de turbine à gaz, Doctoral Thesis, ECP-ONERA, Paris, 1 June 2012.

Dubois J., Smith A.J., " ESA PlasmA RAdiation DatabasE (PARADE). Development history, status, current, developments and future prospects", The 2nd International Workshop on Radiation of High Temperature Gases in Atmospheric Entry, Università La Sapienza, Rome, Italy, 6–8 September 2006.

Dudeck M., Equations linéarisées à l'interface, Rapport du Laboratoire d'aérothermique du CNRS, nos. 76–1, 1976.

Dupays J., Fabignon Y., Orlandi O., et al., " Combustion of aluminium particles in solid rocket motors", ODAS 2000 – ONERA/DLR Aerospace Symposium, Berlin, Germany, 15–16 June 2000.

Dupoirieux F., Tessé L., Avila S., et al., "An optimized reciprocity Monte Carlo method for the calculation of radiative transfer in media of various optical thicknesses", International Journal of Heat and Mass Transfer, vol. 49, pp. 1310–1319, 2006.

Duval R., Transferts radiatifs dans les moteurs à propergol solide, Doctoral Thesis, Ecole centrale de Paris, 2002.

Eggers J., "Nonlinear dynamics and breakup of free-surface flows", Reviews of Modern Physics, vol. 69, p. 865, 1997.

Emmons H.W., "The film combustion of liquid fuel", Zeitschrift für Angewandte Mathematik und Mechanik (ZAMM), vol. 36, pp. 60–71, 1956.

Fabre J., Liné A., "Modeling of two phase slug flow", Annu. Rev. Fluid Mech., vol. 24, pp. 21–46, 1992.

Fernandez V.G., Berthoumieu P., Lavergne G., " Primary Atomization in Water and Kerosene Liquid Sheets At High Pressure", ICLASS 2009, 11th Triennial International Annual Conference on Liquid Atomization and Spray Systems, Vail, Colorado, United States, July 2009.

Fernandez-Pello A.C., "Flame spread modeling", Combustion Science and Tecnology, vol. 39, pp. 119–134, 1984.

Flament C., "Chemical and vibrational non-equilibrium nozzle flow calculation by an implicit upwind method", Proceedings of the 8th GAMM-Conference, Notes on Num. Fluid Mech., vol. 29, pp. 128–137, 1990.

Fortier A., Mécanique des suspensions, Masson, Paris, 1967.

Frackoviak B., Approche expérimentale et simulation numérique des effets d'interactions entre gouttes en évaporation, Doctoral Thesis, ENSAE-ONERA, Toulouse, 26 janvier 2007.

Frenklach M., "On the driving force of PAH production", Symposium (International) on Combustion, vol. 22, no. 1, pp. 1075–1082, 1989.

Frenklach M., Wang H., " Detailed mechanism and modeling of soot particle formation", in H. Bockhorn (ed.), Soot Formation in Combustion: Mechanisms and Models, pp. 165–192, Springer-Verlag, Berlin, 1994.

Gatignol R., "The Faxén formulae for a rigid sphere in an unsteady non-uniform Stokes flow", Journal de Mécanique Théorique et Appliquée, vol. 1, pp. 143–160, 1983.

Gatignol R., Prud'Homme R., Mechanical and thermodynamical modeling of fluid interfaces, Series on Advances in Mathematics for Applied Sciences, World Scientific, vol. 58, 2001.

Germain P., Mécanique des milieux continus, Masson, Paris, 1962.

Ginovska, Bojana. "Computational study of chemical reactions." Online access for everyone, 2007.

Glauz R.D., "Combined subsonic supersonic gas particle flow", ARS Preprint, nos. 1717–61, 1961.

Goldstein S., "The steady flow of viscous fluid past a fixed spherical obstacle at small Reynolds number", Proceedings of the Royal Society A, vol. 123, pp. 225–235, 1929.

Goodridge C.L., Tao Shi W., Hentschel H.G.E., et al., "Viscous effects in droplet ejecting capillary waves", Physical Review E, vol. 56, pp. 472–475, 1997.

Gorokhovski M., Herrmann M., "Modeling primary atomization", Annual Review of Fluid Mechanics, vol. 40, pp. 343–366, 2007.

Gorokhovski M., Saveliev V., "Analyses of Kolmogorov's model of breakup", Physics of Fluids, vol. 15, pp. 184–192, 2003.

Gupta R.N., Yoss J.M., Thompson R.A., et al., " A review of reaction rates and thermodynamic and transport properties for an 11-species air model for chemical and thermal nonequilibrium calculations to 30000 K", NASA-RP-1232, 1990.

Guy A., Collisional-radiative and macroscopic models for the thermochemical relaxation of non-equilibrium hypersonic flows, Doctoral Thesis, Ecole Centrale de Paris, 16 December 2013.

Hanson A.R., Domich G.E., Adams H.S., "Shock tube investigation of the break-up of drops by air blasts", Physics of fluids, vol. 6, pp. 1070–1080, 1963.

Hinze J.O., Turbulence, 2nd edition, McGraw-Hill, New York, pp. 460–471, 1975.

Homann H., Bec J., "Finite-size effects in the dynamics of neutrally buoyant particles in turbulent flow", Journal of Fluid Mechanics, vol. 651, pp. 81–91, 2010.

Hultberg J.A., Soo S.L., "Flow of a gas solid suspension through a nozzle", AIAA Paper, nos. 65–6, 1965.

Jenny M., Dusek J., Bouchet G., "Instabilities and transition of a sphere falling or ascending freely in a Newtonian fluid", Journal of Fluid Mechanics, vol. 508, pp. 201–239, 2004.

Joulain P., "Diffusion Flame Combustion in Oxidizing Flow", Micro-Gravity Science and Technology, vol. 4, pp. 178–191, 1993.

Joumani Y., Transferts radiatifs dans les moteurs à propergol solide, Doctoral Thesis, University of Valenciennes and Hainaut Cambraisis, 2001.

Kaminska A., Szymanski Z., Hoffman J., et al., " Ecoulements ionisés dans les plasmatrons pour la simulation des conditions d'entrée en atmosphères planétaires", in C. Izarra, M. Vallee (eds.), Modélisation: atomes, molécules, plasmas, systèmes dynamiques, Presses Universitaires, Bourges, 23 and 24 May 2013.

Kirsch, Harald [Verfasser]. "Chemical reactions on surfaces - a SFG study / Harald Kirsch." Berlin : Freie Universität Berlin, 2014.

Kiwa, Toshihiko, Tatsuki Kamiya, Taiga Morimoto, Kentaro Fujiwara, Yuki Maeno, Yuki Akiwa, Masahiro Iida, et al. "Imaging of Chemical Reactions Using a Terahertz Chemical Microscope." Photonics 6, no. 1 (January 27, 2019): 10.

Kliegel J.R., "One dimensional flow of a gas particle system", IAS Paper, nos. 60–5, 1960.

Kralchevsky P.A., Nagayama K., Particles at fluid interfaces and membranes, Elsevier, Amsterdam, 2001.

Krülle G., Mayer W., Schley C.A., " Recent advances in H2/O2 high pressure coaxial injector performance analysis", AIAA/SAE/ASME/ASEE 26th Joint Propulsion Conference, Orlando, United States, 16–18 July 1990.

Kuentzmann P., Aérothermochimie des suspensions, Mémoire des sciences physiques, Fasc. 72, Gauthiers-Villars, Paris, 1973.

Lalo M., Atomisation d'un film mince liquide par action combinée des instabilités de Kelvin-Helmholtz et de Faraday. Application aux injecteurs aérodynamiques des turbomachines aéronautiques, Doctoral Thesis, ENSAE-ONERA, Toulouse, 21 December 2006.

Lamb H., Hydrodynamic, Cambridge University Press, 1945.

Lamet J.-M., Transferts radiatifs dans les écoulements hypersoniques de rentrée atmosphérique terrestre, Doctoral Thesis, Ecole Centrale de Paris, 2009.

Landau L., Lifschitz E., Mécanique des fluides, MIR, Moscow, 1971.

Lang R.J., "Ultrasonic atomisation of liquids", The Journal of the Acoustical Society of America, vol. 34, no. 1, 1962.

Lasheras J.C., Villermaux E., Hopfinger E., "Break-up and atomisation of a round water jet by a high-spead annular jet", Journal of Fluid Mechanics, vol. 357, pp. 351–379, 1998.

Law C.K., Sirignano W.A., "Unsteady droplet combustion with droplet heating-II: conduction limit", Combustion and Flame, vol. 28, pp. 17–186, 1977.

Le Moyne L., " Atomisation, pulvérisation et aérosols", Techniques de l'Ingénieur, Référence AF 3622, 2010.

Legendre D., Colin C., Fabre J., et al., "Influence of gravity upon the bubble distribution in a turbulent pipe flow: comparison between numerical simulations and experimental data", Journal de Chimie Physique, vol. 96, pp. 951–957, 1999.

Legros G., Fuentes A., Baillargeat J., et al., "Three-dimensional recomposition of the absorption field inside a non-buoyant sooting diffusion flame", Optical Letters, vol. 30, pp. 3311–3313, 2005.

Legros G., Fuentes A., Rouvreau S., et al., "Transport mechanisms controlling soot production inside a non-buoyant laminar diffusion flame", Proceedings of the Combustion Institute, vol. 32, pp. 2461–2470, 2009.

Legros G., Joulain P., Valenton J.P., et al., "Soot volume fraction measurements in a three-dimensional laminar diffusion flame established in microgravity", Combustion Science and Technology, vol. 178, pp. 813–835, 2006.

Lhuissier H., Nappes, trous, ligaments et gouttes, Doctoral Thesis, University of Provence-Aix-Marseille I, 2011.

Li, Rui, Dacheng Feng, and Feng Zhu. "Quantum chemical study of penicillin: Reactions after acylation." International Journal of Quantum Chemistry 107, no. 10 (2007): 2032–39.

Lignell D. O., Chen J.H., Smith P.J., et al., "The effect of flame structure on soot formation and transport in turbulent non-premixed flames using direct numerical simulation", Combustion and Flame, vol. 151, pp. 2–28, 2007.

Ludwig C.B., Malkmus W., Reardon J.E., et al., Handbook of infrared radiation from combustion gases, NASA Technical Report SP-3080, Washington, 1973.

Lype E.F., "One dimensional analysis of non-isentropic two phase flow", ARS Preprint, nos. 1605–61, 1961.

Machicoane N., Zimmermann R., Fiabane L., et al., "Large sphere motion in a nonhomogeneous turbulent flow", New Journal of Physics, vol. 16, 2014.

Marble F.E., "Dynamics of dusty gases", Ann. Rev. Fluid Mech, vol. 2, pp. 397–446, 1970.

Marble F.E., Broadwell J.E., The coherent flame model for turbulent chemical reactions, Technical Report TRW-9-PU, Project Squid Headquarters, Purdue University, 1977.

Marmottant P., Atomisation d'un courant liquide dans un courant gazeux, Doctoral Thesis, Institut polytechnique de Grenoble, 2001.

Marmottant P., Villermaux E., "Fragmentation of stretched liquid ligaments", Physics of fluids, vol. 16, no. 8, pp. 2732–2741, 2004.

Marmottant P., Villermaux E., "On spray formation", Journal of Fluid Mechanics, vol. 498, pp. 73–111, 2004.

Mauriot Y., Prud'Homme R., Assessment of evaporation equilibrium and stability concerning an acoustically excited drop in combustion products, Comptes Rendus Mécanique, vol. 342, no. 4, pp. 240–253, 2014.

Mauriot Y., Prud'Homme R., High frequency combustion instability activities – Droplet dynamics (Task 3.2), Rapport Technique ONERA, no. 4/12499 DEFA, November 2008.

Maxey M.R., Riley J.J., "Equation of motion for a small rigid sphere in a uniform flow", Physics of Fluids, vol. 26, no. 4, pp. 883–889, 1983.

McGlashan, N. R. "Chemical-looping combustion — a thermodynamic study." Proceedings of the Institution of Mechanical Engineers, Part C: Journal of Mechanical Engineering Science 222, no. 6 (June 1, 2008): 1005–19.

Mehta R.S., Haworh D.C., Modest M.F., "Composition PDF/photon Monte Carlo modeling of moderately sooting turbulent jet flames", Combustion and Flame, vol. 157, no. 4, pp. 982–994, 2010.

Modest M.F., Radiative Heat Transfer, 2nd edition, Academic Press, New York, 2003.

Mugele R.A., Evans H.I., "Droplet size distribution in sprays", Ind. Engng. Chem, vol. 43, no. 4, pp. 1317–24, 1951.

Murrone A., ECLIPPS. Adaptation de la chaîne de calcul SPIREE à la problématique du givrage, Rapport Technique ONERA, no. 7, February 2006.

Nickling D.J., Wilkes J.O., Davidson J.F., "Two phase flow in vertical tubes", Transactions of the Chemical Institute of Chemical Engineering, vol. 40, pp. 61–68, 1962.

Nicoli C., Haldenwang P., " Etude en boucle ouverte de l'action résonnante d'une onde acoustique sur une flamme de brouillard", 19e Congrès Français de Mécanique, Marseille, 24–28 August 2009.

Nicoli C., Haldenwang P., " Flamme de brouillard en présence d'une onde acoustique", Colloque du GDR Micropesanteur Fondamentale et Appliquée, La Villa Clythia - Fréjus, 26, 27 and 28 November 2009.

Nicoli C., Haldenwang P., "A resonant response of self-pulsating spray-flame submitted to acoustic wave", Combustion Science and Technology, vol. 182, nos. 4–6, 2010.

Nicoli C., Haldenwang P., Suard S., "Analysis of pulsating spray-flames propagating in lean two-phase mixtures with unity Lewis number", Combustion and flame, vol. 143, pp. 299–312, 2005.

Nigmatulin R.L., Dynamics of Multiphase Media, Hemisphere Publishing Corporation, 1991.

Olesen M.J., Prediction of drop-size distributions based on ligament breakup, Doctoral Thesis, Queen's University, Kingston, Canada, July 1997.

Omiya, Kazuhiro. "Theoretical Study on Chemical Reactions in Solution." ¬Ný-'Yf[(Kyoto University), 2003.

Oron A., Davis S.H., Bankoff S.G., "Long-scale evolution of thin liquid films", Review of Modern Physics, vol. 69, no. 3, 1997.

Padday J., Pétré J., Rusu C., et al., "The shape stability and breakage of pendant liquid bridges", Journal of Fluid Mechanics, vol. 352, pp. 177–204, 1997.

Padday J., Pitt A., "The stability of axisymmetric menisci", Philosophical Transactions of the Royal Society of London A, vol. 275, pp. 489–528, 1973.

Park C., " On convergence of computation of reacting flows", AIAA paper, 1985.

Park C., "Assessment of Two-Temperature Kinetic Model for Ionizing Air", Journal of Thermophysics, July 1989.

Parker S.H., Suh H.K., Lee C.S., "Effect of cavitating flow on the flow and fuel atomisation characteristics of biodiesel and diesel fuels", Energy and Fuels, vol. 22, pp. 605–613, 2008.

Pascaud S., Vers la simulation aux grandes échelles des écoulements turbulents diphasiques réactifs: application aux foyers aéronautiques, Doctoral Thesis, Institut national polytechnique de Toulouse, 28 June 2006.

Peters N., Turbulent combustion, Cambridge University Press, 2000.

Planchette C., Laurenceau E., Brenn G., "The onset of fragmentation in binary liquid drop collisions", Journal of Fluid Mechanics, vol. 702, pp. 7–25, 2012.

Poinssot T., Veynante D., Theoretical and Numerical Combustion, 2nd edition, Edwards, Flourtown, 2005.

Proudman I., Pearson J.R.A., "Expansions at small Reynolds numbers for the flow past a sphere and a circular cylinder", Journal of Fluid Mechanics, vol. 2, pp. 237–262, 1957.

Prud'Homme R., Analyse théorique du chauffage transitoire et de la réponse dynamique à une oscillation acoustique d'une goutte en évaporation, ONERA final report, June 2008.

Prud'Homme R., Flows of reactive Fluids in Homogeneous Mixtures, ISTE, London, and John Wiley & Sons, New York, 2013.

Prud'Homme R., Flows of Reactive Fluids, FMIA Series, vol. 94, Springer, Berlin, 2010.

Prud'Homme R., Flows of Reactive Fluids, ISTE, London, and John Wiley & Sons, New York, 2012.

Prud'Homme R., Habiballah M., Matuszewski L., et al., "Theoretical analysis of dynamic response of a vaporizing droplet to an acoustic oscillation", Journal of Propulsion and Power, vol. 26, no. 1, pp. 74–83, 2010.

Prud'Homme R., Habiballah M., Nicole A., " On Transient heating and dynamic response to acoustic oscillation of a vaporizing droplet", XXXI Summer School - Conference "Advanced Problems in Mechanic" (APM2003), Saint-Petersburg, Russia, 23 July 2003.

Prud'Homme R., Jamet N., Notions de base sur les équations des brouillards, ONERA technical report, June 1996.

Prud'Homme R., La méthode du maximum d'entropie appliqué à l'atomisation des jets liquides, ONERA technical report, 1998.

Prud'Homme R., Préau S., " Echanges thermiques et phases fluides", ICTEA-HT3-06, Beirut, Lebanon, 31 May to 4 June 2004.

Qureshi N.M., Bourgoin M., Baudet C., et al., "Turbulent tranport of materials particles: an experimental investigation of finite size effects", Physical Review Letters, vol. 99, p. 184502, 2007.

Rangwala A.S., " Flame Spread Analysis using a Variable B-Number", Fire Safety Science-proceedings of the ninth International Symposium, pp. 243–254, 2008.

Rayleigh J.W.S., "On the instability of jets", Proceedings of the London Mathematical Society, vol. 1–10, no. 1, pp. 4–13, 1878.

Raynal L., Instabilité et entraînement à l'interface d'une couche de mélange liquide gaz, Doctoral Thesis, Joseph Fourier University (Grenoble I), 7 October 1997.

Refloch A., Courbet B., Murrone A., et al., "CEDRE Software", Aerospace Laboratory, no. 2, March 2011.

Ritter H., Marraffa L., Ferracina L., et al., " ESA technology roadmaps ongoing and planned developments", 7th European Workshop on Thermal Protection Systems & Hot Structure, 8 April 2013.

Roger M., Coelho P.J., Da Silva C.B., "The influence of the non-resolved scales on thermal radiation in large eddy simulation of turbulent flows: a fundamental study", International Journal of Heat and Mass Transfer, vol. 53, pp. 2897–2907, 2010.

Sagaut P., Large Eddy Simulation for Incompressible Flows, Springer, Berlin, 2006.

Sellens R., Zeng W., " Modeling Spectral Influences in Droplet/Turbulence Interaction", ILASS Americas, 11th Annual Conference on Liquid Atomization and Spray Systems, Sacramento, United States, May 1998.

Senoner J.M., Sanjosé M., Lederlin T., et al., "Eulerian and Lagrangian Large-Eddy Simulations of an evaporating two-phase flow", Comptes Rendus de Mécanique, vol. 337, pp. 458–468, 2009.

Sgro L.A., Filippo A.D., Lanzuolo G., et al., "Characterization of nanoparticles of organic carbon (NOC) produced in rich premixed flames by differential mobility analysis", Proceedings of the Combustion Institute, vol. 31, no. 1, pp. 631–638, 2007.

Sindayihebura D., Pulvérisation ultrasonique de films liquides de faible épaisseur, Doctoral Thesis, Université Catholique de Louvain, 1995.

Situ M., Schetz J.A., “Fragmentation Computational and experimental study of the effect of mass transfer on liquid-jet breakup”, AIAA Journal, vol. 23, pp. 254–261, 1985.

Soo S.L., Multiphase Fluid Dynamics, Science Press and Gower Technical, 1990.

Soubrie T., Prise en compte de l’ionisation et du rayonnement dans la modélisation des écoulements de rentrée terrestre ou martienne, Doctoral Thesis, Ecole nationale supérieure de l’aéronautique et de l’espace, 2006.

Sunderland P.B., Quintiere J.G., Tabaka G.A., et al., “Analysis and measurement of candle flame shapes”, Proceedings of the Combustion Institute, vol. 33, pp. 2489–2496, 2011.

Takahashi F., “ Candle and jet diffusion flames: mechanisms of combustion under gravity and microgravity conditions”, in J. Jarosinski, B. Veyssière, (eds.), Combustion Phenomena, CRC Press, Boca Raton, pp. 170–178, 2009.

Tchen C.M., Mean value and correlation problems connected with the motion of small particles suspended in a turbulent fluid, Doctoral Thesis, University of Delft, 1947.

Tessé L., Dupoirieux F., Taine J., “Monte Carlo modeling of radiative transfer in in a turbulent sooty flame”, International Journal of Heat and Mass Transfer, vol. 47, pp. 555–572, 2004.

Tessé L., Dupoirieux F., Zamuner B., et al., “Radiative transfer in real gases using reciprocal and forward Monte Carlo methods and a correlated-k approach”, International Journal of Heat and Mass Transfer, vol. 45, pp. 2797–2814, 2002.

Tessé L., Lamet J.-M., “Radiative Transfer Modeling Developed at ONERA for Numerical Simulations of Reactive Flows”, Journal Aerospace Lab, no. 2, pp. 1–19, 2011.

Tessé L., Modélisation des transferts radiatifs dans les flammes turbulentes par une méthode de Monte Carlo, Doctoral Thesis, Ecole Centrale de Paris, 2001.

Theofanous T.G., Li G.J., “On the physics of aerobreakup”, Physics of Fluids, vol. 20, pp. 1070–1080, 2008.

Tio K.-K., Linian A., Lasheras J.C., et al, “On the dynamics of buoyant and heavy particles in a periodic Stuart vortex flow”, Journal of Fluid Mechanics, vol. 254, pp. 671–699, 1993.

Torero J., Vietoris T., Legros G., et al., “Estimation of a Total Mass Transfer Number from the Stand-off Distance of a Spreading Flame”, Combustion Science and Technology, vol. 174, pp. 187–203, 2002.

Torero J., Vietoris T., Legros G., et al., "Evaluation d'un nombre de transfert de masse réel d'une flamme ascendante", Journal de Physique IV France, vol. 11, 2001.

Tóth B., Lema M.R., Steelant J., et al., "Assessment of slag accumulation in solid rocket boosters: Part I, single-phase", AIAA Paper, pp. 1–10, 2005.

Trouvé A., Poinsot T., "The evolution equation for the flame surface density in premixed combustion", Journal of fluid mechanics, vol. 278, pp. 1–31, 1994.

Tryggvason G., Scardovelli R., Zaleski S., Direct numerical simulations of gas-liquid multiphase flows, Cambridge University Press, 2011.

Tulegenov, Akyl, and Dina Akbayeva. "Quantum-chemical study of activation of yellow phosphorus by copper(II) chloride." Chemical Bulletin of Kazakh National University, no. 3-4 (December 30, 2016): 4–9.

Uhlmann M., "An immersed boundary method with direct forcing for the simulation of particulate flows", Journal of Computational Physics, vol. 209, pp. 448–476, 2005.

Vallet A., Burluka A.A., Borghi R., "Development of an Eulerian model for the atomization of a liquid jet", Atomization and Sprays, vol. 11, pp. 619–642, 2001.

Veyssiere B., Bozier P., Mécanismes de propagation des flammes dans les mélanges hétérogènes gaz-particules solides, Bilan à 4 ans du GDR 2258 CNES-CNRS: Phénomènes de transport et transitions de Phase en Micropesanteur, Paris, September 2003.

Vietoris T., Joulain P., Torero J., "Experimental characterisation of a laminar diffusion flame in micro-gravity", Journal de Chimie Physique, vol. 96, pp. 1022–1030, 1999.

Villermaux E., "Fragmentation", Annual Review of Fluid Mechanics, vol. 39, pp. 419–446, 2007.

Villermaux E., Bossa B., "Single-drop fragmentation determine size distribution of raindrops", Nature Physics, vol. 5, pp. 696–702, September 2009.

Villermaux E., Marmottant E., "Fragmentation of stretched liquid ligaments", Physics of Fluids, vol. 16, pp. 2732–2741, 2004.

Wallis G.B., One-Dimensional Two-Phase Flow, McGraw-Hill, New York, 1969.

Wang H., Du D.X., Sung C.J., et al., "Experiments and numerical simulation on soot formation in opposed-jet ethylene diffusion flames", Symposium (International) on Combustion, vol. 26, no. 2, pp. 2359–2368, 1996.

Wang H.Y., Cordeiro P., Joulain P., "Direct numerical simulation of a laminar diffusion flame established over a horizontal flat plate in micro-gravity environment", Journal de Physique IV France, vol. 11, pp. 331–338, 2001.

Wang H.Y., Joulain P., El-Rabii H., " Modélisation numérique et étude expérimentale sur la propagation d'une flamme de diffusion sous gravité réduite avec du transfert radiatif", Colloque du GDR Micropesanteur Fondamentale et appliquée, Fréjus, 26–28 November 2007.

Williams F.A., Combustion Theory, 2nd edition, Benjamin Cummings, Menlo Park, 1985.

Xu F., El-Leathy M., Kim C.H., et al., "Soot surface oxidation in hydrocarbon/air diffusion flames at atmospheric pressure", Combustion and Flame, vol. 132, nos. 1–2, pp. 43–57, 2003.

Yecko P., Zaleski S., "Transient growth in two-phase mixing layers", Journal of Fluid Mechanics, vol. 528, pp. 43–52, 2005.

Young J.B., "The fundamental equations of gas-droplet multiphase flow", International Journal of Multiphase Flow, vol. 21, no. 2, pp. 175–191, 1995.

Zao B., Yang Z., Li Z., et al., "Particle size distribution function of incipient soot in laminar premixed ethylene flames: effect of flame temperature", Proceedings of the Combustion Institute, vol. 30, no. 1, pp. 1641–1448, 2005.

INDEX

www.ingramcontent.com/pod-product-compliance
Lightning Source LLC
LaVergne TN
LVHW062226200926
848484LV00006B/1482